The Impact
of Geology
on the United States

The Impact
of Geology
on the United States

A Reference Guide to Benefits and Hazards

ANGUS M. GUNN

Foreword by BRUCE BABBITT

Greenwood Press
Westport, Connecticut • London

Library of Congress Cataloging-in-Publication Data

Gunn, Angus M. (Angus Macleod), 1920–
    The impact of geology on the United States : a reference guide to benefits and hazards /
    Angus M. Gunn ; foreword by Bruce Babbitt.
        p.   cm.
    Includes bibliographical references (p.   ).
    ISBN 0–313–31444–6 (alk. paper)
    1. Environmental geology—United States.   I. Title.   II. Index.
QE38.G86   2001
363.34'0973—dc21   00–061710

British Library Cataloguing in Publication Data is available.

Library of Congress Catalog Card Number: 00–061710
ISBN: 0–313–31444–6

First published in 2001

Greenwood Press, 88 Post Road West, Westport, CT 06881
An imprint of Greenwood Publishing Group, Inc.
www.greenwood.com

Printed in the United States of America

The paper used in this book complies with the
Permanent Paper Standard issued by the National
Information Standards Organization (Z39.48–1984).

10 9 8 7 6 5 4 3 2 1

To Ruth, my wife,
a true friend for 40 years

CONTENTS

FOREWORD: AMERICAN RESTORATION

The current generation of Americans is not just approaching but is already crossing the threshold into an entirely new era, a third great environmental thrust that I will call "The Restoration Movement." The first era was the conservation movement of Teddy Roosevelt, which created America's great parks, wildlife refuges, and national forests. It lay the legal foundation. The second generation, that of Rachel Carson, saw our air and water and soil being polluted by modern industrial society and helped pass legislation that brought us clean water and clean air.

The current generation is now awakening to a new and larger vision, to the possibilities that we can use our laws not just to stop decline but to reverse it; not just to preserve the isolated parts but to protect and reconnect whole landscapes and entire watersheds; not just to fence off the local greenway or trickling neighborhood stream but to unite them with the great national parks and the wide ocean-bound rivers. Across the landscape this generation is opening a new chapter, an entirely new era of conservation called American Restoration.

It's hard work, back-breaking work, but we are willing to do it because the process feels good in our hands and our spirits. It feels good to sit in, eat on, and to drive. It also looks good and gives us an aesthetic pleasure when we see it in our living rooms or driveways. It is something we can do with our spouse, our sons and daughters, and so become a stronger and prouder family because of our work.

What is happening, however, in this generation, goes beyond our home and neighborhood to involve the entire community.

In the course of my travels through every region of the nation I saw Americans crossing the threshold from prevention to restoration. They are building on the current framework of laws, getting results beyond the expectations of the legislators who wrote our laws back in the 1970s. Our federal laws do not require communities to restore their local waters, but in order to restore their local waters, those communities require our laws. Americans say to me: We have done this and we are proud of what we have done, but we can do more. We have stopped the decline of our rivers, the erosion of our soil, the disappearance of our open land, but we can go further. We will back their efforts to restore every single watershed in America. We will help them reclaim their heritage.

The environment is not just a fixed point in time or some place outdoors or even an issue to be tackled. It is rather a tradition that endures only through our labor, an opportunity that lasts as long as we fight for it. It is part of our collective heritage, to be passed on like a torch, a job that brings us all together under a common purpose.

Bruce Babbitt, former Secretary of the Interior

(from a recent report, here by permission of the Department of the Interior)

ACKNOWLEDGMENTS

I am greatly indebted to many individuals and institutions for their help: Bruce Babbitt, former Secretary of the Interior, for permission to use his statement on the environment; the U.S. Geological Survey, for its extensive support via publications and personal communications; the Geological Society of America, for permission to reproduce some of its maps and drawings; the Institute of Professional Geologists, for permission to reproduce some of its slides; Professor Vladimir E. Romanovsky, of the University of Alaska, for alerting me to the serious effects of global warming in that state; and to state geologists throughout the nation, for their detailed, invaluable assistance. I would also like to single out Dr. Robert Fickies, Geological Survey, State of New York, and Michael Bograd, Office of Geology, State of Mississippi, who have been exceptionally generous with their time and resources; last, but by no means least, Emily M. Birch of Greenwood Press, whose enthusiasm, efficiency, and thoughtfulness enabled me to complete the manuscript much sooner than I had expected.

INTRODUCTION: GEOLOGY AND U.S. ENVIRONMENTS

REGIONAL FRAMEWORK AND SCOPE OF VOLUME

Geological regions define the framework for this book. This is appropriate since it is geological activities beneath the surface of the earth that give shape over time to the present form of the surface. Our task is to examine the kinds of rocks found in the different regions, look at activities that are presently reshaping them, both geological and climatic, then investigate the impact of these forces on our environment along with ways of minimizing their destructive effects.

In addition to an overall examination of geological processes, representative states will be selected to illustrate hazards and geologic features found over larger areas. In this way, the resources of states can be drawn on for further in-depth study. I want to avoid the generalized type of book in which examples of geological processes are taken from widely scattered locations, leaving the reader without a clear focus for follow-up investigations.

To assist in further study, four appendixes are included at the end of the book: a geological time chart showing the different ages and their names; a list of U.S. Census Bureau divisions and regions, to aid in finding statistical data; the names of all the active and potentially active volcanoes in the United States; and finally, comparisons between metric and imperial measures. In addition, bibliographies are provided at the end of this introduction and after each chapter to direct further study, and a comprehensive list of reference volumes follows the glossary.

This book will provide a readable and readily available resource for geological students in advanced high school and introductory-level university courses and will meet the needs of those studying physical geography or earth science. While geologic regions form the areal framework for the text, the climatic elements shaping our environment will also be examined. Thus, contemporary issues such as the following will be addressed: earthquakes and volcanism in the Cordillera, subsidence in the Appalachians, water resources on the Great Plains, and exceptional weather along the Atlantic Coastal Plain.

Early American traditions in geography and earth science included coverage of geology as an integral component of area studies. In recent decades, however, largely due to the growing emphasis on human aspects of these fields, geological content has been reduced to a level of superficiality that demeans its content and deprives a student of a great key to environmental and economic problems. Now, with the huge advances being made in our understanding of plate tectonics and its influence on day-to-day life, there is added justification for a renewed approach to the study of geological processes and their impact on the environment.

The land surface of the United States is a mix of many forms and different materials, the end result of geological forces acting over long periods of time and modified by wind, rain, and temperature. They provide scenery and recreational settings for us, the beauty of the east coast beaches, the awe-inspiring depths of the Grand Canyon, and numerous wilderness settings like Yellowstone National Park. They are sources of such national wealth as minerals and oil, and their watercourses, even today, are key transportation routes.

The processes that shape this landscape are ongoing and frequently catch us by surprise when unexpected things happen. It is still more troubling when general patterns change. Global warming is just one example, affecting the behavior of weather systems all across the country; hurricanes and tornadoes become more powerful and cause greater damage, rainfall increases, and low-lying areas experience flooding. Damage to the ozone layer is another, caused by CFCs, chlorofluorocarbon gases, which increase the risk of damaging radiation. Fortunately, there is good news alongside the bad: CFCs are on the decline. In the United States they dropped to one third of their former level in the course of the 1990s. For the carbon dioxide emissions, the ones that cause global warming, intense efforts are under way to reduce the amounts being released into the atmosphere.

Whether geological and climatic processes continue as before or whether they change, the hazards associated with them are a continuing concern. We have little chance to predict destructive events because many of them are random occurrences. Earthquakes and volcanic eruptions are particularly difficult to predict, yet their ability to disrupt and destroy is great. The Alaskan quake of 1964, the San Fernando quake in 1971, and the Loma Prieta quake in northern California in 1989 (Figure I.1) are three earthquakes from the second half of the twentieth century that caused loss of both life and property along the Pacific margins of California and Alaska. All three were unexpected.

Hazards are not restricted to what might be termed *natural causes*. Humans are responsible for some of the destruction that occurs. Warnings of hundred-

Figure I.1. An example of tragedy from design, the upper deck of the Nimitz Freeway Interstate 880 overpass in Oakland, California, which collapsed during the 1989 Loma Prieta earthquake, the worst to hit central California since 1906. Design weaknesses may cause failure again in future earthquakes unless corrective measures are taken.

Figure I.2. The power of Hurricane Hugo's winds and waves is evident in the condition of buildings, both temporary and permanent, on both sides of the water. Notice the debris in the water and on land. This photograph was taken in 1989.

year flood risks—that is to say, the likelihood of a major flood happening approximately every 100 years—are sometimes ignored in the interests of short-term gain or just because, in our fast-paced lifestyle, a hundred years is too big a time frame to comprehend. Again and again, often just a few years after a devastating flood, developers are able to persuade authorities to allow new sub-divisions on the same flood plains. They know they may not be around to face responsibility when the next big flood comes.

The sheer growth of population in vulnerable areas is yet another human-induced danger. More and more people have been moving to the southeast coasts, to the Carolinas and Florida, so whenever a hurricane strikes, its dev-astation is always greater than an identical storm at an earlier time simply be-cause there are more people in the path of the storm. Hugo was the worst tropical cyclone on record when it struck South Carolina in 1989 and caused $6 billion in property losses and the deaths of 29 people (Figure I.2). Three years later, Hurricane Andrew hit an area south of Miami and created a new record for destruction, far more property losses than resulted from Hugo and twice as many deaths (Figure I.3).

ENVIRONMENT

While this book deals only with environmental issues arising from geologic and climatic processes, we need to remember that environmental problems range

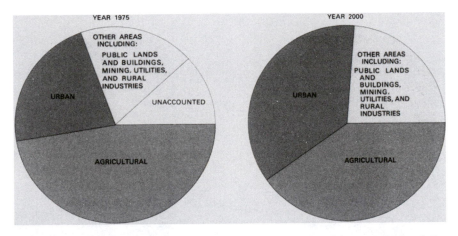

Figure I.3. Trends in the distribution of flood losses. The total increase in the dollar value of damage from $3.4 billion to $4.3 billion is only an increase of 26 percent over a 10-year period, but the shift from agriculture to urban, buildings, industries, and utilities means that more people are affected and the risk to humans therefore greater.

over a much bigger field than these: Too much specialization in crops leads to a depletion of the gene pool with a consequent higher risk of diseases; toxic accumulations in our freshwater cause illnesses and add the expense of purifying systems; serious ecological damage to our parks and wilderness areas occurs as more and more people and businesses use them; several accidents have raised concern over the safety of nuclear installations and the storage of nuclear waste materials; the probability of earth being seriously damaged by an asteroid is now being investigated with renewed care; finally, the increase of violence in society creates concerns about crime and terrorism.

Because of all these concerns, environmental issues are now receiving more and more attention from all levels of society. Environmental movements are major players in all that is happening. They are numerous, and they are concerned about all of the above and others. The rapid depletion of old-growth forests in the Pacific Northwest, for example, had environmental protection movements lock horns with forest companies at the beginning of the 1990s. Fortunately, both sides were able to settle their disputes through mutual compromises. Things are not always so easily resolved. Remedies are sometimes proposed by environmentalists that are completely unacceptable to the companies concerned.

In the fall of 1999, in a desert area about 100 kilometers east of Los Angeles, the Fish and Wildlife Service stopped a half-billion-dollar development that would have brought homes and shopping centers to the area. Why? Because this small region of sand dunes is the only habitat of the tiny Delhi Sands flower-loving fly, an orange-brown insect the size of a pin, the only fly ever to make the list of endangered species. This case is not unique. There have been other

Figure I.4. Waste dumps from underground lead and zinc mines in northeast Oklahoma. These materials spoil the appearance of the landscape, and as they are eroded by rain, their toxic materials are carried into the groundwater systems.

events of the same kind in the past. The big difference now is the much higher priority being given to preserving existing habitats.

Mining

Mining is an industry that is not well understood by many of its critics. It operates underground for the most part, and the things that get publicized about it are the toxic materials in the waste that is left on the surface. One illustration is enough to see how public attitudes are formed. Cyanide may be part of the waste at a gold mine, so our first reaction is to demand the enactment of the strictest possible laws for waste disposal. The importance of the minerals being mined tends to be eclipsed in our concern over contamination of land and water (Figure I.4). Thus when public bodies make laws about land use, mining is often a low priority; parks, wilderness areas, and territory for endangered birds, animals, or even insects get the most attention.

Mining is a high-risk occupation. Large sums of money can be invested in a mine and production undertaken only to discover that the value of the mineral can be lost overnight, as it were, through discovery of a lower-cost deposit of the same kind elsewhere or a new finding that the health risks associated with the mineral render it unusable. Asbestos lost its value when it was found to be a major health hazard. Yet, despite the risks, minerals are more than ever essential to our way of life. We need some rare minerals like platinum if we are to reduce emissions from cars, one of the very strong demands of environmentalists. We need the mineral silicon for computers, the instruments that reduce our consumption of raw materials and enable us to cut back on the use of our cars—another goal of environmentalists.

Recycling is now being taken up by the mining industry for good reasons, and this development might allay the negative attitudes to which it has been subjected. Many of the ores being mined are more complex than they were formerly and often of lower grade, therefore more costly to process. Use of scrap metals is one way of reducing costs. In the production of steel this trend is particularly evident. Over 60 percent is being manufactured from recycled ferrous scrap. The ready supply of old steel and the development of new furnace technology have made it easy to take advantage of metals that would otherwise be thrown away.

Another incentive for recycling metals relates to landfill, the traditional location for dumping unwanted things. The costs of using this source are rising rapidly because fewer and fewer sites are still available. As a result, wastes that might not have been economically viable for recycling a short time ago are now attractive because landfill costs would be much higher. In spite of this fact, and the growing insistence from the public for it, regulations remain as a hindrance. Stainless steel scrap destined for recycling is sometimes classified as hazardous waste because of its content of nickel and chromium, two of the varied alloys conserved during stainless steel recycling. Even the category of hazardous waste is a problem because it raises the costs of recycling.

Public perception is yet another hurdle to be overcome in the quest for optimum use of recycling. There is a widespread feeling that products made from recycled materials are inferior. For example, military aircraft specifications ban the use of metals made from scrap for rotating engine parts. Virgin material must be used. These requirements come from a lack of confidence on the part of aircraft designers that recycled materials will stand up to the rigors of an operating engine. The challenge for researchers in the mineral industry is to prove that recycled materials are fully capable of meeting the physical and structural requirements of military engines.

Water

The preservation of adequate supplies of clean water is always near the top of any country's highest priorities, and we will see as we go along the different approaches being taken to improve water quality throughout the nation. The question that is more difficult to tackle than improving water quality is this: Where do we go for freshwater when supplies run out through inefficient usage or climate-induced droughts? Throughout the Cordillera and in large sections of the Great Plains, water withdrawals both from surface waters and from groundwater are high, and as we will see in later chapters, many problems arise when groundwater is excessively exploited.

In Antelope Valley, California, home of Edwards Air Force Base and the Space Shuttle, natural groundwater and seasonal supplies of surface water may not be adequate to meet future demands. This is an arid area, and groundwater withdrawals have already caused irreversible compaction of clay and silt layers in the aquifer from which water is pumped. Land has already subsided and caused substantial damage to runways, roads, wells, pipelines, and buildings generally. Surface cracks in the land surface have also appeared, some of them as large as 400 meters long and 4 meters deep.

One widely discussed solution to freshwater shortages lies north of the forty-ninth parallel where surplus is the norm. Canada has substantially more surface water available annually than is available in the United States, yet its total population is only one tenth of that in the United States. Mexico is much less endowed. With approximately one third of the U.S. population, it has only one third of the U.S. water total. As a result, many proposals were advanced for the diversion of Canadian water southward, especially after the United States–Canada Trade Agreement was signed.

Some proposed shipping billions of liters of water by supertankers from Newfoundland and the central areas of British Columbia's coast to the United States. Opposition to these plans was strong on the grounds that local habitats would be permanently destroyed if so much water were to be withdrawn. Those in favor pointed out that only surplus water is shipped, leaving lakes and rivers with the same volumes as before. Opposition continued as different spokespersons mentioned the possibility of water levels being affected by future climatic changes and the enormous difficulty of changing arrangements once parts of the United States become dependent on water from Canada.

Most ambitious of all is the North American Water and Power Alliance (NAWPA) plan, a design that not only serves the needs of the United States but also meets Canada's demand for a better distribution of its water resources. The prairie lands of Alberta and Saskatchewan, for instance, are greatly in need of freshwater for agriculture. NAWPA proposes a massive lake in the Rocky Mountain Trench, a valley running north and south in the Rockies, and a series of pipelines or canals leading from this lake both eastward and southward. Dams on major rivers such as the Fraser and Columbia that have their sources in the Rockies would supply water for the big lake. This plan is now many years old but still very much alive.

CLIMATE AND GEOLOGY

Air and water are major agents in geological processes. If we look at human tragedies around the world, we find that for every disaster involving 100 or

Figure I.5. Farther inland, in spite of the protection that a wooded area like the one here normally provides, Hurricane Hugo's power was also felt here. Few of these homes were unaffected. This is a 1989 photograph.

more deaths, 55 percent are caused by flooding, that is, too much water, or else by drought or too little water. Twenty percent are the result of wind storms. In every case, terrain conditions, the uneven land surface created by geological action, aid in these destructive activities. We use the term *climate* to describe the combined action of air and water and the different temperature regimes within which they act.

Climate may be defined as average conditions in the atmosphere in contrast to the day-to-day weather; so, by this definition, we have climatic regions, places on the earth's surface that have predictable features from month to month and from year to year. Average features of climate are not easy terms to grasp because there are all kinds of exceptional climatic events that occur, and any one of them can be so enormously destructive that it makes little sense to say that they come, on the average, so many times a year and affect this part and that part of the country. I am thinking of hurricanes, tornadoes, and floods when I say "exceptional" events. Their numbers and intensities can often be predicted, but they affect a particular part of the country so infrequently—sometimes as little as once in five years—that they can hardly be taken as part of average conditions (Figure I.5).

El Niño is yet another exceptional event but less frequent than any of the three listed above. It is caused by a periodic warming of the Pacific Ocean's surface waters along the coast of Ecuador. The name is derived from the Spanish word for "The Christ Child" because the warming is often first observed around

Christmastime. Because of the continuous interaction that takes place between the surface of the earth and the atmosphere, El Niño immediately affects the air masses over North America, providing it, among other things, with one or two years of warmer-than-average winters.

The climatic variations listed above, and also the appearance of a massive volcano once in 50 years, blasting gases and volcanic particles around the world and so reducing the amount of insolation that reaches the earth's surface, are all short-time changes in the bigger picture of climatic uncertainties. If we think about a very much longer time frame, say 50,000 years, then an altogether different set of uncertainties appears: Both the path of the earth as it orbits the sun and the orientation of the earth's axis relative to the earth's plane of orbit change dramatically.

The orbit changes slowly over approximately 50,000 years from elliptical, the pattern we regard as normal because it happens to be the present pattern, to circular, then through the same stretch of time back to elliptical. Parallel to the orbital changes are the earth's axial movements through a cycle of angular changes twice as fast as the orbital ones. This is quite different from the climatic variations that take place over a few years, and their influences are correspondingly much greater. The series of ice ages covering half of North America during the last million years of earth's history is one outstanding example of their effects.

In between the short-term climatic variations and those extending over thousands of years is the global warming trend of the past 30 years, which is receiving a lot of attention at the present time. This one is blamed on human rather than natural activity, on the release of increasing amounts of carbon dioxide into the atmosphere. Evidence has been amassed in recent years to show that industrialization with its accompanying high consumption of fossil fuels is the culprit in this rapid increase. The net result is a greenhouse effect; that is to say, atmospheric temperature is increasing at an alarming rate because the carbon dioxide prevents long-wave radiation from escaping into the outer reaches of the atmosphere. Pressure is now increasing worldwide for a reduction in fossil fuel consumption.

GEOLOGICAL HAZARDS

Geological and associated hazards are studied in order to devise methodologies for limiting losses when the next earthquake or flood occurs. I mentioned the growth of population in vulnerable areas as one human-induced hazard. Alongside this must be placed the growth of cities, by far the biggest transformation ever to occur in human settlement. Large numbers of people are found within the confines of a small area, and a complex and costly infrastructure is built to serve them. An earthquake in such a location will cause damage far beyond anything that a similar hazard could cause in a rural setting.

In many communities, geologists, planners, and political leaders are working together to achieve a balance between the conflicting goals of economic development and public safety. One costly solution is to exclude homes and other kinds of buildings from places where the hazard risk is high. If the cost of this is too high in economic losses, and this is exactly the kind of conflict local authorities face, then a compromise is to reduce the density of buildings through land-use zoning bylaws. Good design at the beginning is yet another way of protecting buildings against severe strain and so limiting losses.

Historical records document many examples of significant losses because geological processes were not adequately considered. These experiences are warnings of future damage, but they are not always given serious consideration. For example, the 1989 Loma Prieta earthquake in Santa Cruz County caused many of the same kinds of landslides that were noted after quakes in 1865, 1868, and 1906. Geological information can be used for safer land development not only in earthquake-prone California but in all parts of the United States. There are many examples of disasters where prior attention to geological data would have greatly reduced damage. There are also examples of the wise use of geological information.

Loma Prieta Elementary School was built on top of fault lines near the epicenter of the 1989 quake, but no prior geological assessment was required when it was built, so no one knew of the danger that lurked underneath. When, about a year before the earthquake, portable classrooms were needed to cope with the growing number of students, the office of the state architect required a geologic investigation. It was at that point that the foundation problem was discovered. A decision was made to abandon the site and move to a new location. The earthquake intervened before the move was made, and there was considerable damage; fortunately, it struck when the school was empty.

Love Creek canyon in Santa Cruz County experienced a huge debris flow during heavy rains in 1982, killing 10 people and destroying nine homes (Figure I.6). Follow-up studies revealed the instability of the canyon and neighboring territory with the result that several homes were removed and the area evacuated. The Love Creek community had been developed over the years with cabin-style second homes long before geologic studies were required, and people continued to improve and convert these homes right up to the time of the debris disaster.

The Marina District of San Francisco is a stretch of unconsolidated material close to the Bay. During the 1906 earthquake a near-liquefaction condition was created by the shaking, and many homes were destroyed. In 1915 this same part of the city became the site of a national exhibition. A neighboring lagoon was filled with sand to provide additional space. Later, still more fill was added, and the site became a residential district. The lessons of 1906 were not learned, and when Loma Prieta struck, liquefaction and amplified ground shaking caused much greater damage than in other parts of the city. Four people died, and 70 homes were either totally destroyed or rendered unsafe.

A similar neglect of the dangers of unconsolidated material led to the destruction of 75 homes in Alaska during the 1964 quake. Homes had been built on a

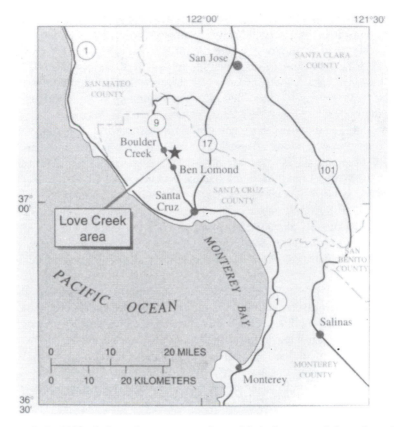

Figure I.6. In 1982, during winter storms, a huge debris flow roared down Love Creek canyon, burying 10 people and destroying nine homes. Post-disaster studies revealed an adjacent area with 25 homes that was quite unstable, so the county of Santa Cruz had to move the homes to a safer location.

bluff under which lay a bed of clay long known by experts to be a hazard in the event of an earthquake. The site was a very desirable one because of the excellent view of the ocean. As the quake shook the ground, the whole bluff disintegrated and tumbled in a series of blocks into the water. Three lives were lost, and all of the subdivision's roads and utilities were destroyed.

Two examples of subsidence illustrate a different kind of neglect. In Texas a residential area near Houston had been gradually sinking over a period of years as a result of groundwater withdrawal, and the level of the land was moving closer and closer to sea level. There was no recognition of the dangers of subsidence when these homes were built. By 1979 they were so close to sea level that residents had to move into upper floors several times a year as high tides flooded the ground floors. Hurricane Alicia hit the area in 1983 and completely destroyed the whole subdivision.

A subsidence of a different kind occurred in Florida in 1981 when a housing complex built over limestone rock collapsed. Limestone is a rock that dissolves in water over time, so open spaces are created as underground streams eat away at the underlying rock. At some point the ground above these open spaces gets too thin and the surface layers give way. In this instance near Orlando, the sinkhole swallowed part of a swimming pool, two buildings, a home, and several automobiles. Total losses amounted to $2 million. As in the case of Texas, knowledge of the geology of the area could have prevented this tragedy.

The long-standing neglect of geological hazards is now being attended to in many jurisdictions. In Utah in the 1970s and 1980s, studies identified the risk of a serious earthquake on a fault that runs right through Salt Lake City (Figure I.7). Despite this, the city allowed a large apartment complex to be built directly on this fault. A group of citizens protested, and they succeeded in persuading the city to buy the rest of the land around the fault and establish a park, appropriately named "Faultline Park," both to prevent further development and to remind citizens of the continuing danger. Similar preventive action was taken in one location in California. To avoid a recurrence of destructive landslides, authorities there decided that soil and geologic studies must be conducted wherever construction is proposed for slopes steeper than 15 degrees.

Natural Hazards

Having stressed the place of human responsibility and intervention in geological events, it is essential to point out that all of these happenings are perfectly normal and natural, necessary parts of earth's creative activity. Without them life on our planet would not be possible. Furthermore, many of the geological or climatic events that happen and we deem to be bad have, alongside their destructive effects, good outcomes: Floods that destroy farm buildings fertilize the farmland; earthquakes that destroy homes ignite forest fires that are essential to the preservation of an ecosystem. These natural events are neither benevolent nor malevolent. They are neutral. It is humans who transform the environment into resources or hazards.

Everything we see today—mountains, oceans, river valleys, and so on—were created by geological actions, all of them accompanied by earthquakes and volcanism. For almost all of the time in which these events took place, we humans were not here, so there were no hazards, only beneficial, massive earthquakes and gigantic volcanoes, all busy at work getting the planet ready for life as we know it now. In the course of this ongoing geological activity, many valuable substances appear at or near the earth's surface. Some of these, such as asbestos and coal, are both valuable and hazardous at the same time, and the challenge for us is how best to use them. Their natural state tells us nothing more than that they exist.

The long history of earth's development has its parallel today. Nothing is static. Everything is dynamic. The only reason it does not appear to be so is the

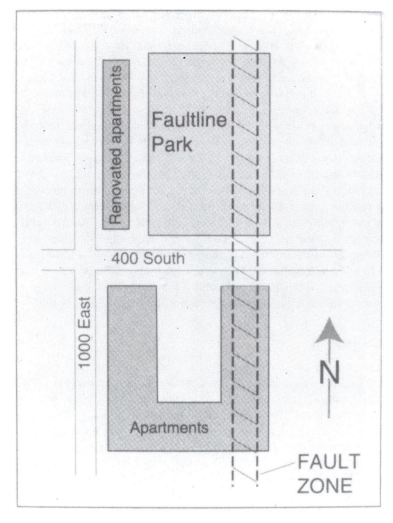

Figure I.7. The Wasatch Fault, recently active, runs through Salt Lake City and is a potential source of a damaging earthquake. A group of citizens, watching a large apartment complex being built directly on this fault, persuaded the city to purchase some adjacent land for a park as a demonstration of good and bad land-use practice.

slowness of the changes that are occurring. Mountains appear to be fixed all through our lifetimes, and references from the past suggest that they were always like that. Today, with the instruments at our disposal—such as the Global Positioning System, which can measure earth movements of a centimeter a year—we know that change is everywhere. Natural events with which our grandparents were familiar, and against whose destructive potential they made preparations, may no longer be relevant. Things may be worse now, or they may be better, and people may hanker or not for the so-called good old days.

Climate change is one assured event that will affect people within their own lifetimes. Climate is a measure of average weather conditions, and it can be stable over 5 years, 10 years, or even longer stretches of time, but on average, we can expect to see big changes in levels of precipitation and temperature in any one area within the lifetime of any one person. There are also the more recent discoveries of short-term shifts in climate. Whenever these things happen, there are unexpected floods or droughts, increases or decreases in hurricanes or tornadoes, and sometimes unexpected clusters in any one of these events. We discovered recently that earthquakes do not always come, on average, after known intervals of time. Unfortunately, they also sometimes come in clusters, dislocating predictions.

A number of geologic changes can be predicted due to the tools at our disposal nowadays. We know, for example, that sea levels on the shores of Connecticut and Massachusetts will rise by at least a meter over the next 100 years, and we know how much of that amount will occur within the next 10 years. The reasons for the rise in sea level date back to the last ice age when the weight of ice depressed the land, and it has been slowly recovering ever since. The problem with this kind of data is a human one: What action should I take today? To ignore what is happening means that thousands of homes will face destruction unless action is taken. We need to continue our research efforts in order to understand better what natural geological forces are at work in our midst.

REFERENCES FOR FURTHER STUDY

Burton, I., R. W. Kates, and G. F. White. *The Environment as Hazard*. New York: Oxford University Press, 1978.

Clark, W. C. *The Human Dimensions of Global Environmental Change*. Washington, D.C.: National Academic Press, 1989.

Foster, H. D. *Disaster Planning: The Preservation of Life and Property*. New York: Springer Verlag, 1980.

Foster, H. D. *Health, Disease, and the Environment*. London: Belhaven Press, 1992.

Glickman, T. S. *Study of Trends in Disasters, 1945–1986*. Washington, D.C.: Resources for the Future, 1991.

Graedel, T. E., and P. J. Crutzen. *Atmosphere, Climate, and Change*. New York: Scientific American Books, 1995.

Houghton, J. *Global Warming*. Oxford: Lion Publishing, 1994.

Tarbuck, E. J. *Earth Sciences*. 8th ed. Upper Saddle River, NJ: Prentice-Hall, 1997.

White, G. F., ed. *Natural Hazards, Local, National, Global*. New York: Oxford University Press, 1974.

1

ABOUT GEOLOGY

Geology is all about time, and it's all about new ways of thinking about time. When a geologist says that something is recent, he does not mean last week or last year or even a hundred years ago. He is not even thinking about events 10,000 years ago. Rather, he is thinking in millions of years or tens of millions of years. We have to think as if real life were like the speeded-up photography we sometimes see on television—a season's life of a plant in a fraction of a second, a map showing the steady growth of the world's population over 2,000 years displayed in a few seconds, or the development of the last ice age shown as it began, came to full coverage, and died away, all presented in a one-minute series of maps to represent 100,000 years.

The time frame we need to describe the earth is nothing less than the total history of the planet, from its faint origins over 4 billion years ago right up to the present. This enormous stretch of time is divided up into eras, periods, and epochs, and you can see them all at a glance in Appendix A. Rocks that are between 5 and, say, 65 million years old fall into the period we call Tertiary. Those younger than 5 million are in the period named Quaternary, and within Quaternary there are smaller units, epochs, such as Holocene, for rocks that may be only 10,000 years old. Tertiary and Quaternary together are grouped into a bigger unit called the Cenozoic era. At the other end of the scale is the Precambrian era, covering the time span from 600 to 4,000 million years ago.

The crust of the earth, the part we see every day, has a density that is less than the parts deeper down inside the earth. This must be so since lighter rocks first rose to the surface of the earth just because they were less dense than other rocks, just as ice rises to the surface of water because it is less dense than water. The density of the crust is easy to calculate with a few representative rocks, but how can we calculate the density of rocks deep inside the earth? Experience of seismic waves from earthquakes is the answer. They travel faster through less dense rocks as well as through those that are rigid or under great pressure and slower through dense ones. Through this technique we now know that the center of the earth is four times the crust's density, and twice that of the mantle, the region between earth's center and its surface. Some geologists conclude that the earth's center must therefore be composed mostly of iron.

ROCKS

It is important to know exactly what we mean by *rocks* because the kinds of rocks in place in any given situation make a big difference to the outcomes when a steep hillside receives a heavy rainfall or is shaken by an earthquake. In all, there are three families of rocks. *Igneous* ones are those that were once molten. They come from inside the earth and usually contain crystals that develop as they cool. *Metamorphic* is the name given to any rock that was subjected to intense heat or pressure, forcing it to recrystallize. *Sedimentary* rocks are the ones that form at the surface of the earth as a result of erosion or solution. These actions give rise to sediments that in time harden into sedimentary rock.

A rock is defined as a coherent mass of solid, not living matter forming part of our earth. By *coherent mass* I mean that the particles of the rock are bonded together into a solid form. A tree cannot be a rock since it is part of living matter, and a collection of loose grains of sand cannot be a rock because they are not consolidated into a firm mass. However, dead plant or animal matter can in time be compressed into rocks; coal and coral are examples. When the surface of the earth is exposed to heat, rain, or ice, rocks break down, and day by day tiny pieces are carried away by wind or water to accumulate as sediment at some lower elevation. Over the long stretches of geological time, these tiny particles solidify into sedimentary rock.

Sedimentary rock accounts for only a small percentage of the earth's crust, but this statistic belies its importance. An examination of the rocks at the earth's surface reveals that most of the outcrops are sedimentary. They constitute a thin, perhaps discontinuous layer on the uppermost part of the crust. This is what we would expect because sediment, the material from which sedimentary rocks are formed, accumulates on the surface. It is from sedimentary rocks that we are able to reconstruct earth's history because only there will we find fossils of former life forms or clues to past environments. Furthermore, many of our natural resources, including coal, petroleum, and natural gas, are found in sedimentary rock.

If the sediments that formed the sedimentary rock happen to have accumulated on the ocean floor close to the edge of a continent, the rock may be carried down into the interior of the earth by subduction as the spreading ocean crust sinks down beneath the continent. There, in the hot interior, higher temperatures and much greater pressures change sedimentary into metamorphic rock. If the degree of temperature or pressure is moderate, the new rock will look like the

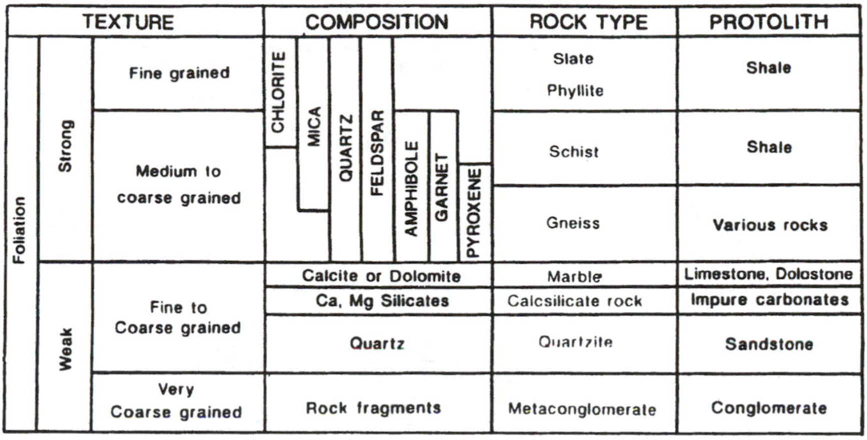

TEXTURE		COMPOSITION	ROCK TYPE	PROTOLITH
Strong	Fine grained	CHLORITE / MICA / QUARTZ / FELDSPAR / AMPHIBOLE / GARNET / PYROXENE	Slate / Phyllite	Shale
Strong	Medium to coarse grained		Schist	Shale
Strong	Medium to coarse grained		Gneiss	Various rocks
Weak	Fine to Coarse grained	Calcite or Dolomite	Marble	Limestone, Dolostone
Weak	Fine to Coarse grained	Ca, Mg Silicates	Calcsilicate rock	Impure carbonates
Weak	Fine to Coarse grained	Quartz	Quartzite	Sandstone
Weak	Very Coarse grained	Rock fragments	Metaconglomerate	Conglomerate

(Left side vertical label: Foliation)

Figure 1.1. An arrangement for naming metamorphic rocks. The left side describes the texture of the rock, whether the minerals are tightly bound together or are weak, and the degree of coarseness of the rock particles. On the right, the source rock that has been metamorphosed is listed against the resultant rock. The middle column lists the amounts of different minerals for each rock listed in the third column.

parent material. Slate is a good example. It is the sedimentary rock shale after metamorphosis, and in appearance it is difficult to distinguish it from shale on the basis of appearance. Slate, because heat realigned its minerals in a layered manner, is easily split into thin slabs and is therefore popular as floor or roof tiles. Slate's color may be black, red, or green, depending on its mineral content.

Metamorphism takes many forms and creates various outcomes: It can increase the density of a rock, grow larger crystals, generate new minerals, and as we saw in the case of slate, rearrange the mineral grains into layered patterns (Figure 1.1). *Schist* is a descriptive term for rocks like slate that have been metamorphosed into layered patterns. It deals only with the texture, not the composition, of a rock. Mica schist is an example. *Gneiss* is another generic term, one that describes elongated patterns. The most common minerals in gneisses are quartz and feldspar. Marble is quite different from schists and gneisses. It has no layered pattern but, rather, is coarse-grained, composed of large calcite crystals that form from the smaller grains in the limestone parent rock. Because of its color and softness, marble is a popular building stone. White marble is a popular choice for monuments or statues. Impurities in the parent limestone provide a range of other colors—pink, gray, green, and black.

Having mentioned *minerals*, we need to define what they are. In a sentence, they are the building blocks that make rocks. Granite is a rock but it is made up of different quantities of several minerals. Quartz, hornblende, and feldspar are three of them. The geological definition of minerals is fourfold—naturally occurring, inorganic, solid, and with a definite chemical structure. There are thousands of minerals, each one with its own unique chemical structure, that is

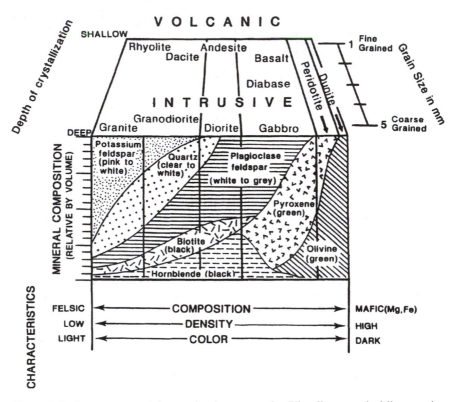

Figure 1.2. An arrangement for naming igneous rocks. Visualize a vertical line running through the diagram to represent the composition of the rock in question. Granite, for example, ranges in composition from 75 percent potassium feldspar plus some other minerals (left side) to only 25 percent potassium feldspar plus some other minerals (right side).

to say, a unique structure made up of one or more of the known 100 elements. Gold and sulfur happen to be minerals that are made up of one element only, but that is exceptional; most are composed of a mixture of elements. The two most abundant elements in earth's surface rocks are oxygen and silicon, and the two most common classes of minerals are silicates (a combination of silicon and oxygen) and carbonates (a combination of carbon and oxygen).

The third kind of rock is igneous, so called after the Latin word for "fire." In some ways it may be considered the primary type of rock because it is by far the most common in the composition of the earth as a whole. It appears on the surface as a result of molten rock, usually called magma, pushing upward through openings in the surface, especially at the midocean ridges, then cooling and solidifying. Different words are used to describe igneous rocks, depending on where they are and how they are behaving (Figure 1.2). If magma cools underground, it is known as plutonic rock; if it flows out on to the surface from a volcano, it is lava; and when lava solidifies, it is called volcanic rock.

All three types of rock are well able to withstand a great deal of shaking from an earthquake, provided care was taken at an earlier stage to avoid construction of buildings on them at faults in the earth's crust. It is at these points that displacements occur when a shaking takes place, moving surfaces either vertically or horizontally and causing breaches in roads, pipelines, or foundations. Faults are numerous and ubiquitous, the end result of millions of years of earth movements. A good deal of work has been done at identifying their locations, but large numbers of them have never reached the surface and are therefore unknown until something happens. The quest goes on continually to find and map new ones.

Unconsolidated surfaces, that is to say, surfaces on which the particles of rock have not yet cemented into a solid mass, pose different problems. Soils with layers of wet sands that lie a meter or two below the surface, a common condition in the Pacific Northwest, are a particular hazard when an earthquake strikes and shaking follows. Water is forced upward, the sand grains separate, and the whole upper soil layer turns into a kind of slurry that behaves like quicksand, sucking buildings down below the surface. The process has appropriately been called *liquefaction*. Fine silt from former riverbeds appears, at first glance, to be a strong substance, well able to support bridges or highways and withstand shaking. Under heavy rain, this type of ground cover can also experience liquefaction. Landfill areas and steep slopes are additional unconsolidated surfaces where ground collapse can occur and with it the destruction of buildings and roadways.

Igneous Intrusions

Not all of the magma that rises up from the interior of the earth erupts onto the surface. Often it solidifies at depth, and we use the phrase *intrusive igneous* or *plutonic rock* for situations like that. One of the important differences between intrusive and extrusive igneous rock is the size of the mineral particles. Extrusive rocks have very small mineral structures. The rates at which different minerals and rocks crystallize explain this important distinction. Basaltic magma begins to crystallize near the surface at 1,200 degrees Celsius, and for granitic magma, it is 1,100 degrees. At depths of, say, 10 kilometers, these figures increase by about 20 degrees because of the pressure. At the same time, minerals crystallize at temperatures that are different from the body of magma—metal-rich ones at higher temperatures and the potassium-rich feldspars at lower temperatures.

Granite, being much more viscous than basalt, is less easily extruded onto the surface, so it tends to stop rising at a point a few kilometers underground where surrounding rocks are cold enough to block further movement upward. Most granites are like that. Millions of years after they have cooled and solidified at the deeper level, they are exposed by erosion to form the massive batholiths we see today. The general structure of these batholiths is coarse-grained. They have larger minerals than those in rocks that reach the surface before crystallizing. One effect of these massive granite intrusions near the surface is that neighboring rock is metamorphosed as a result of exposure to so much heat.

One special form of intrusive igneous rock is the dike. This is where a pulse of magma is injected upward into a vertical fissure. When the magma freezes, a dike is formed. These are vital for the water supply of Hawaii because they serve as impermeable strata to store rainwater. The reason they solidify quickly in fissures is because small quantities of magma ascend through a large volume of cooler rock. Generally, dikes form parallel to one another. Alongside them, but running at right angles, are sills, dikes that form at right angles to the vertical dikes.

EROSION AND DEPOSITION

The earth's surface is never perfectly flat but, rather, consists of slopes at all degrees of steepness. Rocks and soil are therefore always exposed to the influence of gravity, sometimes hardly perceptible over years, at other times extreme as rocks and debris crash down from a cliff. The pull of gravity is aided by the abrasive actions of wind, water, and ice, as well as the disintegrating power of temperature. The last mentioned can be an effective erosive agent even in very dry climates where there is little or no water or ice. By day the rock heats and expands, then contracts at night. Repeated actions of this kind over time cause rock to break apart into smaller and smaller fragments.

Weathering is the first stage of erosion on surface rocks. Water freezing in cracks and crevices splits rocks into smaller pieces; plant roots can have the same effect as they work their way into spaces between rocks; and chemical action such as the influence of acid rain can readily wear away the outer layers of certain minerals. In public buildings that have stood for some time, it is often easy to see the pitted marks on rocks that succumbed to chemical erosion as well as those that were composed of more resistant minerals. Chemical erosion is very powerful, and it comes from many sources, not just acid rain. Water can leach powerful chemicals from rocks. Their actions on other rocks can then increase as other agents create new surfaces on which they can work.

Some agents of erosion are so strong that things happen quickly, and gravity becomes the dominant influence. A glacier can pluck large stones from its bed and carry them to the melt point where, provided the stream is fast enough, they can be transported long distances downslope. Glaciers can also hold large stones in their lower layers and use them to erode other rocks and to deepen their valley floors. Steep cliffs, such as those along the Californian coast that are subject to pounding waves, can be undermined by the waves as layers of sand are removed at the water line. Blocks of debris break off at the top, and the cliff gradually recedes.

The most common modes of transportation for the rock fragments created by weathering are rivers and wind. Once the fragments are small enough, gravity and small rivulets move them into the larger streams that carry them to their next destination lower down. Water's carrying power is related more to its speed than its volume. If the speed is doubled, its ability to carry large boulders may be increased three or four times. While wind can only carry very small fragments, much smaller than those borne by water, it carries them at such high speeds that they can literally sandblast objects in their path.

As rivers reach their lowest level, they slow down and begin to deposit their sediments on their delta regions at the rivers' mouths. On both sides of the river is a large area covered by muds and silts from flood episodes in the past when the main stream was unable to contain the floodwaters in its channel. While the flood is abating on these occasions, the water on the floodplain is stagnant long enough for the suspended load of rock particles to settle out. Over time the floodplain receives successive layers of fine-grained sands and muds, providing a foundation for the best agricultural land in the nation. Under flood conditions, more sediments are deposited immediately next to the river rather than farther away from it, leading to the formation of raised banks, or levees. Those of the Mississippi are good examples.

Many of the drier regions in the western United States have a different experience of river erosion and transportation when unexpected weather conditions bring heavy rain. River valleys that had not seen water for most of the year are suddenly flooded, usually by flash floods that race down the slopes fast enough to carry along quite large boulders. At the bottom of the slope, all the sediment and debris are dispersed in a fan-shaped heap of rubble and vegetation. At the same time, landslides are triggered on the sides of the high ground because vegetation cover is sparse. In some parts of Colorado and Utah, during one of these heavy rainstorms, the extra weight of saturated soils was enough to trigger block collapse of some steep slopes.

The volcanic mountains of Hawaii experience a very different kind of sedimentary deposit. Each one of these volcanic peaks goes through a period of sinking deeper and deeper into the ocean as the rock of which it is composed cools and therefore becomes heavier and heavier. On the shorelines of these mountains, coral reefs are constructed. The sea polyps that construct them can only live close to the surface of the ocean, so as the volcanic peak sinks, new generations of these sea creatures build new reefs in order to be close to the surface. Over long periods of time the reefs mount higher and higher. Finally, when the volcano is completely below sea level, there is a circular coral reef, a sedimentary limestone rock, and a lagoon within it.

NATURAL RESOURCES

Geological processes are the key to understanding and finding all of the inorganic raw materials we need to sustain a modern lifestyle. They include coal, petroleum and natural gas, and a range of metals like gold, silver, and copper. We are always talking about finding alternative renewable energy sources that we can afford like solar, wind, waves, tides, and fusion, but for various reasons, these are unable to meet all our needs at the present time. We conclude that most of our energy will continue to come from nonrenewable fossil fuels. In addition to our energy needs, we use many metals that are nonrenewable, so we need to conserve and, where possible, recycle all of these resources simply because they are nonrenewable. They took millions of years to form; we do not want to deplete them in less than a hundred years.

Oil and natural gas have the same origin and are usually found together. The raw material from which they come is dead organic matter such as microorganisms and plants. If enough of this kind of dead organic material collects in a place where there is little oxygen, so that decay is delayed, then the stage is set for a possible future oil well, but a lot of things have to happen along the way before that point is reached. Heat, pressure, and chemical transformation all have to work on the organic matter, and this process takes millions of years.

By the time the plant material has been covered by successive layers of sediment and other organic matter and is buried to a depth of about six kilometers, its temperature reaches 180 degrees Celsius. At this temperature chains of carbon molecules bond with hydrogen to form petroleum; as heat increases further, these chains of oil change to natural gas, but the extra amount of heat is critical. Too much and everything is ruined. It is easy to wonder how oil and gas ever develop. So many stages in their development happen by chance. The oil is under pressure when it distills out of the transformed raw material, and if there is any opening, it will rise to the surface.

Seeps of oil on the surface have been known for a long time. Once its value was recognized, search was not focused on these sites because it was obvious that if oil could reach the surface in this way, it could already have oozed away. So the search was on for big underground reservoirs with no leaks. The most common trap for oil proved to be where the crust was folded, allowing oil to be trapped in the tops of the folds. Most of the big oil reservoirs of the world are found in these folds at varying depths from one to five kilometers. The many random factors that lead to the final, recoverable oil are so numerous that only one fiftieth of the world's organic debris ends up as oil.

Coal has a similar origin to oil, mainly from land plants that were abundant over 200 million years ago and accumulated in such enormous quantities that today we have huge resources available on continent after continent. The same process of protection provided by avoiding contact with oxygen and the same sequence of compression under new deposits were operative here as well. Again, like oil and gas, the various kinds of coal formed according to degrees of compression and temperature. Partly because it is a solid rather than a liquid, coal is more stable. Apart from those that have been mined, all the seams of coal that were formed millions of years ago still exist.

When it comes to the mining of metals, there are huge value differences between them, something that does not apply to any significant degree with fossil fuels. The ore containing a particular metal must contain a sufficient percentage of the metal to justify mining it. For iron, which is abundant, the ore must contain 50 percent iron; for platinum, at the other end of the spectrum, mining is economically worthwhile if only one ten millionth of the ore is platinum. In between there is a whole range of values, and sometimes there are ores that contain several metals, leading to an average value that would make the mine economically viable.

Natural resources include many others besides these more well-known ones. Carbonate rocks, that is to say, limestone or marble, are widely used in the construction industry for highways and buildings. Marble is metamorphosed limestone. Clay minerals are extremely fine-grained sediments that settled at the bottom of ancient glacial lakes and are now used in the making of bricks, pottery, and stoneware dishes. Granite is traditionally the preferred choice for large buildings, but in recent years, it has been replaced with less costly substitutes. Slate, metamorphosed shale, is an important resource for flagstones, flooring tile, and roofing.

A SCIENTIFIC REVOLUTION

It is rare that a field of knowledge as big as geology should have its entire history of theories turned on its head, as it were, yet that is what happened because of new findings in the 1950s and 1960s. For centuries, thousands of geological books and reports appeared, describing just about every possible aspect of this subject but, at the same time, raising many questions regarding the causes of the phenomena described. Then, in the later 1960s, a consensus developed within the scholarly community on a brand-new theory, one that deals with most of the unanswered questions from the past and totally revolutionizes traditional ways of thinking about geology. The full meaning of this new theory is still being worked out, but its comprehensiveness is not.

There had been many attempts to find a unifying theory. Most fields of science develop in this way. New data come to light, and ideas are advanced; but without the tools or opportunities to test them, they get pushed aside sooner or later. At one point, out of frustration over the lack of a powerful theory, one famous scientist asserted that geology's intellectual powers were no higher than those needed to collect stamps. The data that kept accumulating were indeed difficult to comprehend: Why are the rocks of Madagascar similar to those found in India when these areas are separated by the Indian Ocean? Why is the top of Mount Everest made of rocks that contain fossils, remains of creatures that once lived in the sea? What explains vestiges of glaciers in the Sahara and tropical jungles in Alaska?

James Hutton's work at the end of the nineteenth century introduced the idea of geological time. It was a major step toward a better understanding of the earth. Geological processes for the first time were envisioned in millions instead of thousands of years. Hutton saw the huge gulf between human and geological time, the difference between the clock in the church tower and the one in the mountain. Alfred Wegener, in 1915, introduced another idea, one just as revolutionary as Hutton's: The continents of the world were once joined together in a single mass that he named "Pangea"—meaning "all earth." In his mind, continents moved about on the ocean like icebergs. His idea was true in principle, as was proved in the 1960s, but it was too radical for his time.

By the 1930s scientists in South Africa and Switzerland had developed Wegener's ideas and were able to establish beyond any doubt the reality of continental drift, but finding an explanation for it was the problem. How could continents drift? Those who advocated the idea of huge land masses sliding over ocean floors had to contend with a majority of geophysicists who regarded the whole idea as impossible. The two sides were deadlocked right up to the years following World War II when, under the military requirements of the cold war with its need to detect Soviet submarines, the United States launched a massive program of ocean floor research. For the first time ever, the 70 percent of the earth's surface that is covered by water began to yield a succession of secrets.

The most unexpected discovery was a series of connected submarine ridges running all the way around the earth. They lay approximately midway between North America and Europe and between Africa and South America on the Atlantic side, and off center in the Pacific, where they were much closer to North and South America. The ridges were huge mountain ranges, rising from the ocean floor to heights of 3,000 meters. From their peaks downward ran steep, narrow valleys along the full length of the mountain chain. This was by far the longest mountain range and the longest valley anywhere on the planet; yet this was the first time anyone knew of their existence.

Other discoveries followed. Existing theories about the origin of the earth held that it had cooled from a primordial molten ball, so it was expected that beneath the silt and other sediments on the ocean floor the oldest rocks of all would be found, safely protected from the erosion that is always changing the surface features of the land. What researchers did find were rocks no older than 150 million years—young indeed by geological standards. How could this be? One scholar suggested that perhaps magma was rising up from inside the earth through the underwater mountain valleys and covering the ocean floor with young rock. It was already well known that these valleys were the sites of numerous earthquakes.

It was not long before the idea of moving plates took shape. Other discoveries strengthened the idea: Young ocean floor was identified all across the Atlantic on both sides of the underground mountains, so the magma from the valleys must have traveled great distances; at the eastern edges of the Pacific Ocean, deep trenches were discovered, and these, too, were known as familiar locations for massive earthquakes, much stronger than the ones in midocean, some of them extending down as far as 700 kilometers underground. Could this young rock on the ocean floor be a huge conveyor belt of volcanic rock, originating in midocean mountain ridges, traveling toward the continents, sinking down beneath them, melting as it reaches the deeper parts of the mantle, and reappearing again at the ridges millions of years later?

The clincher, confirming the theory of the conveyor belt, came from studies of the earth's magnetic field. The whole earth behaves like a single magnet. Wherever a compass is placed on its surface, the needle points in the same direction. Geologists knew that molten lava, as soon as it cools, is magnetized by the earth so that its magnetic field is thereafter permanently aligned in the same way as that of the earth at the time it cooled. But the earth's magnetism is not always the same. It is thought that, on average, it completely reverses itself every half-million years. North becomes south, and south becomes north. Scientists knew how to date very old rocks using the radioactive decay method, so here at last was an ideal method both of finding out the ages of the rocks on the ocean floor and, at the same time, confirming what was suspected about magnetic reversals.

A ship sailing from side to side across the Atlantic, at right angles to the mountain ridge in the middle, with magnetometers trailing behind and scanning the sea floor, recorded at regular intervals the magnetic orientation of the volcanic rock as well as the distance traveled. When the results were mapped later, there were a series of alternating strips of ocean floor, one pointing northward, the next southward, the one after that north again, and so on. This zebra-like arrangement of magnetic orientation in the volcanic rock on the sea floor was matched by an identical pattern on the other side of the midocean ridge. The distance from the midocean ridge of the first strip, and its magnetic orientation, was exactly the same as the first strip on the other side. It turned out to be the same for all subsequent strips.

The story was now complete. Molten lava was being erupted in the center of the ridge, pushed sideways, and cooled as more rock arrived. As it cooled, it adopted the magnetic orientation of the earth at that time. On islands that are integral parts of the moving volcanic crust, the speed at which the new crust spreads across the ocean from the ridge can nowadays be measured very accurately by such tools as the Global Positioning System. Thus we can mark off the progress of volcanic rock on the ocean floor with a map scaled in millions of years, where each change in magnetic orientation is represented by a specific distance and so many millions of years.

The general picture is of an earth broken into moving pieces of lithosphere, seven large ones, each consisting of both oceanic and continental portions, and several smaller ones. Each plate is about 80 kilometers thick and has a shallow part that deforms by elastic bending and a deeper part that yields plastically. Deeper still is a viscous layer on which the entire plate slides. The plates interact mostly at their edges, as motions between two parts of a spherical surface. This fact of interaction on a spherical surface is confusing when we try to visualize action while looking at flat maps, but it does explain the patterns we find on the surface. Most plates are moving relative to others all the time. Rates of movement range from 1 to 13 centimeters per year.

REFERENCES FOR FURTHER STUDY

Barton, Robert. *The Oceans*. London: Aldus Books, 1980.
Bolt, Bruce A. *Earthquakes and Geological Discovery*. New York: Scientific American Library, 1993.
Bolt, Bruce A. *Inside the Earth*. San Francisco: W. H. Freeman, 1982.
Hallam, A. *Great Geological Controversies*. Oxford: Oxford University Press, 1983.
Motz, Lloyd, ed. *Rediscovery of the Earth*. New York: Van Nostrand Reinhold, 1979.
Sullivan, Walter. *Continents in Motion*. New York: McGraw-Hill, 1974.
Thorndike, Joseph J., Jr., ed. *Mysteries of the Deep*. New York: American Heritage Publishing, 1980.

2

PLATE TECTONICS

For most of human history there was little interest in the idea that the earth might be very old. After all, mountains and valleys seemed to be little changed from year to year or, for that matter, from century to century. Many people, perhaps the vast majority, thought that the earth was only as old as recorded human history, about 5,000 or 6,000 years. One of the first to question this view of the past was a Scot named James Hutton. He was very interested in the natural world, and in 1788, his inquiries into the origins of rocks led to a revolutionary breakthrough in the way people thought about earth's history.

On the east coast of Scotland he found horizontal layers of sandstones on top of beds of shales and sandstones all nearly vertical and some folded back on themselves. To Hutton's mind all this represented a break between two quite distinct episodes in the accumulation of sedimentary rocks. He concluded that rivers must have eroded an ancient landscape, shifting fragments of rock as sediment down to the sea and then compacting them into sedimentary rock. Later these rock layers were uplifted and twisted around until they were vertical, and then a new cycle of erosion began. This process continued until the land surface was worn down and the accumulated sediments were pressured into sedimentary rocks. Hutton believed these cycles of erosion, deposition, and uplift could be repeated indefinitely.

He knew that these sedimentary rocks were formed by processes such as river erosion that can be observed today and that these processes are always extremely slow. He concluded that the processes of uplift must also be extremely slow. Furthermore, it was obvious to him that all the processes he had observed occurred more than once. The implications were shattering. The things he had observed in the rocks seemed to demand an immense, almost limitless amount of time for their formation. This was the beginning of the concept of geological time, and to Hutton it suggested, as he put it, "no vestige of a beginning, no prospect of an end." Throughout the whole of the nineteenth century geologists sought to fill out the details of Hutton's extraordinary find, seeking both the sequences and ages of the different rock layers. Precise measurements of age did not come for a century after Hutton's historic discovery.

Of equal significance to the discovery that earth's history stretched back an unimaginable amount of time was the awareness that the forces at work in the ancient past were the same as the ones operating today. This idea, known as *uniformitarianism*, stood in sharp contrast to many current notions of a catastrophic past in which mountains and oceans took shape, almost instantly as it were, as a result of some massive upheaval. Hutton's discovery has sometimes been compared with that of Copernicus, who showed that the sun is the center of our solar system, not the earth, as had been thought. Copernicus gave us a new awareness of the vastness of space. Hutton did the same regarding the enormity of time.

Even before Hutton's discovery, scholars were speculating on another concept, that of the mobility of continents and oceans. In 1800, German geographer Alexander von Humboldt insisted that the lands on both sides of the Atlantic Ocean had once been joined. The traditional view held that the earth was a stable mass of continents and oceans. The shakings and eruptions that occurred from time to time were seen by most as random events, unrelated to the overall layout. Even the best of scientific thinkers believed that the earth had changed little over long periods of time. It once was a molten mass, and as it cooled and contracted, its crust crumpled into the wrinkled mass of mountains we see today. Throughout the nineteenth century, scholar after scholar published discourses on drifting continents, only to see their ideas rejected because of these prevailing views.

The strongest challenge to these views came from German scientist Alfred Wegener. He argued that the continents were not always where they are now, and in 1915, he published a book entitled *The Origin of the Continents and Oceans* in which he proposed a theory of continental drift. He showed that the continents were once joined together into a single land mass and that over geological time they had drifted apart to their present positions. The gaps that opened up between the continents became our present oceans. Supposedly the idea of floating continents first came to him during a visit to Greenland, where he saw icebergs breaking away from land and moving across the water.

The evidence that Wegener advanced in support of his theory included the existence of similar ancient forms of life on pairs of widely separated continents that are now so far apart that their flora and fauna could not have been in contact across such distances. Other similarities in rock formations and, in such cases as Africa and South America, the ease with which the land masses fit together added further support to his theory. If, said Wegener, the continents were assembled together into the original "Pangea," a Greek word meaning "all land," then all the distributions as indicated by the geological records would be appro-

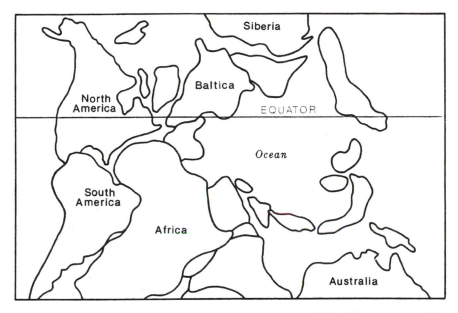

Figure 2.1. Pangea as it probably appeared between 200 and 250 million years ago. The shapes of three of the present-day continents can readily be identified. This cycle of change from one supercontinent to the map of today occurs as oceans develop. A similar supercontinent will probably again form after another 200 million years.

priately connected (Figure 2.1). He illustrated his idea by comparing it to a newspaper that has been torn into bits. As the pieces are brought together again, the original paper can be identified by seeing the lines of print running smoothly across the whole.

Wegener backed up his theory with measurements of rock densities on land and at the ocean floor. He found that the granite rocks of the continents had a density of 2.7 grams per cubic centimeter, whereas the ocean crust had one of 2.95 grams per cubic centimeter. He felt that there was balance isostatically between the lighter continents and the heavier ocean floor. He thought that the ocean floor was the top of the earth's mantle and that the continents could move within it. In his lifetime and for many years after his death in 1930, Wegener's theories remained controversial. Nevertheless, evidence continued to accumulate that continents had not always been in their present locations. There are deposits, for example, in now-tropical India that are of the same age as 300-million-year-old fossilized tropical reefs in lands that are now in the Arctic.

Many scholars were convinced by such indicators as these that the continents had moved from their present positions. Geologists from South Africa and Britain took up Wegener's cause after his death, but they were consistently baffled by the problem of "how." What force was powerful enough to move continents thousands of kilometers across an ocean? Geophysicists regarded the concept of continents as rafts sliding over ocean floors as impossible. As recently as the end of World War II the views of those opposed to Wegener's thesis still held sway. Nothing had been proved to change that outlook, but it was not long before an accident of history led to quite new thinking. Political actions proved to be a great help to science, as we will see when we consider the nature of the ocean's floor.

EARTH'S CRUST AND MANTLE

Before we continue, we should look at the form and behavior of the earth's outermost layers, the crust and mantle. These rocks in the surface layers of the earth take up different amounts of space, some as much as 100 kilometers, and they are the lighter rocks of the crust. This is why they are at the surface. They have worked their way to the top, just as ice will rise to the surface in water because it is lighter, or less dense. Continental crust rises higher than oceanic crust because it is lighter. The floor of the ocean, as we shall see later, is made up of basalt, a very heavy rock compared with the materials that comprise land and mountains. The continental crust is not only lighter than oceanic; it is much thicker, sometimes as much as 10 times thicker.

The interior of the earth is extremely hot, as is well known to miners who have to work a thousand meters or more underground. There is a steady increase of temperature as you go down from the surface, in many places as much as one degree Celsius for every 15 meters of depth. This well-known fact means that heat is flowing upward from the interior to the surface, with the volume being high in volcanic and geothermal areas and low in stable continental places that have not experienced geological disturbances for millions of years. It is the necessity of heat being able to escape from the interior of the earth that explains the convection-type movement of rock in the upper mantle, an action that is part of plate movements.

At depths of 50 kilometers earth's temperature is about 800 degrees Celsius. This is not quite as hot as molten magma, which can be as high as 1,500 degrees Celsius, but it does mean that rock at 50 kilometers down might melt. Temperature, however, is not the only factor to be considered. Pressure also increases with depth, and rock that would melt at the surface is sufficiently pressured at depth that it will not liquefy. We know that this is true because certain earthquake waves that can only move through solids still travel almost 3,000 kilometers deep into the earth. In fact, the lithosphere is solid for considerable distances down, more on the continental parts than on the oceanic.

This lithosphere, or crust, extends as a rigid mass, as deep as 150 kilometers beneath the continents and less than 10 under the oceans. It seems to float on a layer of mantle rock beneath the crust, an area known as the *asthenosphere*, meaning weak. Both the oceanic and continental crusts sink down into this part of the mantle so that they are in a sense floating on it. You could use the term *plastic* for the asthenosphere because it allows rock to move within it. The technical term for such is *isostasy*, and it applies almost everywhere. For instance, after the ice ages ended, land gradually rose higher out of the mantle as the extra weight was removed.

Though technically made of solid rock, some of the asthenosphere circulates very slowly in a pattern called solid-state convection, in which heat is distributed from the deeper parts of the mantle to the base of the lithosphere. This escape of heat from the earth's interior is essential. Small amounts of it are conducted directly through the lithosphere, but much more surfaces through places in mid-ocean. This deeper zone of mantle on which the lithosphere, as it were, floats is therefore somewhat viscous, and it extends downward as far as 700 kilometers.

THE OCEAN FLOOR

The political accident to which I referred came soon after 1945. Cold war tensions were increasing, and one major threat lay in the Soviet Union's nuclear submarines. These ships operated over long stretches of ocean, so if their whereabouts were to be tracked, new knowledge of the sea floor was required. Existing ideas about the appearance and character of the seabed were quite vague, and there was reason to be concerned about their accuracy. Intensive research projects were launched by the United States, and first discoveries revealed an ocean floor quite different from what anyone expected.

Some awareness of the uneven nature of the ocean floor was familiar as far back as the sixteenth century, and in later centuries, additional information was added. The traditional process of finding depths was the tedious process of "swinging the lead," taking soundings by lowering a weighted line over the ship's edge. The sonar methods developed during World War II constituted a large advance. Acoustic echo sounders, which sent high-pitched sounds into the deep and analyzed their echoes, proved to be valuable in delineating the topography of large tracts of ocean floor.

Improved technology continued to appear throughout the 1950s and 1960s, and with these, geologists were able to map still more extensive areas of ocean. Submersibles, small submarines that were specially designed to penetrate deeper than had ever been done before, made it possible for the first time to see some of the deepest parts of the sea. A high degree of precision was reached in mapping the terrain of the ocean floor, matching the accuracy possible on land where visual techniques are the norm.

Deep trenches near the margins of continents, reaching down several kilometers, and relatively shallow midocean ridges were two of the big discoveries that emerged. The Aleutian Trench on the southern margins of Alaska, for example, is more than seven kilometers deep. On the continental side, it is very steep and toward the ocean less so. From these features came the discovery that the ocean floor must be much younger than the continents. If, as had been thought earlier, the seabed dated back to the earliest ages of the earth and was

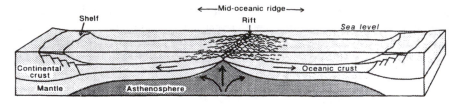

Figure 2.2. The ongoing action at the midocean ridges. Magma rises to the surface, moves sideways, then is pushed farther along as fresh magma erupts. Finally, this ocean crust reaches a continent and either subducts or, as in the case of the Atlantic ridge, pushes the North American continent westward.

at least as old as the oldest parts of the continents, then the elevated areas would long ago have been eroded down to a flat surface and be covered with huge deposits of sediment.

The midocean ridge was found to be a continuous undersea mountain range running all around the earth and coinciding with the places where earthquakes are frequent. On closer inspection it was found that this ridge had a valley down its center for much of its length, indicating a sideways movement of the ocean floor. Once again, older views of earth's history were in conflict with these new discoveries. Here was a mountain area that was being pulled apart, not compressed, as the old ideas of mountains had maintained. The conviction was steadily growing, based on numerous observations, that the valley in the ridge was the place where magma was rising to the surface from deep within the mantle.

It was not long before the concept of gigantic ocean floor plates began to take shape. Magma erupting at these midocean ridges was continually creating fresh underwater mountains, standing higher than the ocean floor because molten rock is lighter than rock that has cooled and solidified. The magma, as it is being thrust upward and outward, cools and forms new ocean crust, which then moves away from the ridge. It travels along the ocean bottom at a rate of a few centimeters per year, a fast pace geologically, and when it reaches a barrier such as a continent, it sinks down under it because its volcanic rock is heavier than that of the land. As it moves down into the deep hot mantle, it becomes molten rock again. Sea floor spreading like this became the key that had eluded Wegener and others all their lives. Here was the answer to how continents could move (Figure 2.2). If a sea is expanding, as clearly is happening, how can continents stand still?

Deep down in the asthenosphere the ocean crust that had melted joined with the convection currents already there, to reappear millions of years later at the midocean ridge. Thus the whole process of sea floor spreading behaved like a huge conveyor belt, the surface part of an equally big convection system within the asthenosphere. A brand-new theory of continental drift had now been established. The big difference now was that the drift was occurring in midocean.

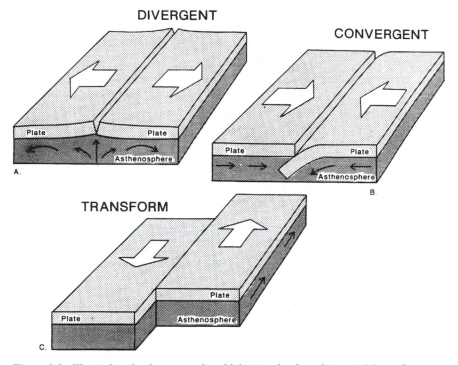

Figure 2.3. Illustrating the three ways in which tectonic plates interact: (A) moving apart as happens in midocean ridges; (B) subducts—the principal action along the west coast of North and South America; or (C) transform movement as two plates slide past each other—occurring in the San Andreas Fault and in southeast Alaska.

The continents do not wander over the floors of oceans but instead are being ferried from place to place on huge barges. These barges are the gigantic tectonic plates formed by the continents and the oceanic crusts to which they are attached (Figure 2.3).

Submersibles went down into the valleys within the midocean ridges in order to further investigate the nature of the eruptions that were taking place. This was a new kind of exploration, comparable to the first voyages across the Atlantic. The kind of submarine needed for the three or four kilometers of descent into the depths was only available in two or three countries. It had to be strong enough to withstand the high pressures yet light enough to be easily handled because the mode of travel was free fall guided by an echo-sounder. Once the submarine was close to the bottom, weights were released to slow down the descent; from there on, it moved around on its battery-powered motors. Travel time to the bottom was usually two hours.

From inside their tiny submarines, they were able to illuminate the landscape of that pitch-dark zone. In the Atlantic Ridge they found black pillows and boulders of black basalt scattered among smaller fragments of broken stones and long tubes of twisted lava, all remnants of recent eruptions. Black smokers were a new reality for the explorers, plumes of muddy water at temperatures of 350 degrees Celsius, gushing out of the sea floor. These were caused by cold seawater entering the crust through cracks, getting heated by hot rock, and then escaping to the surface. On the opposite side of the North American continent, in the valleys of the East Pacific Rise, they made new discoveries, all kinds of life, clams, and crabs clustering by the hot vents.

Iceland is a special case of a midocean ridge. It is above sea level. It happens to lie right in the middle of the Mid-Atlantic Ridge and forms part of it. The cliffs that run north and south across the island are part of the valley in the middle of the ridge, and we now know that they are slowly moving apart, as we would expect. The small amount of magma that wells up in Iceland is only a tiny part of the vast amount that is rising along the rest of the ridge. The frequent earthquakes and volcanoes with which the island is familiar are typical of what is happening now in ridges all around the world. We should note that Iceland, like some other countries, learned to make use of its geothermal sources of power, a valuable locational legacy.

Terranes

From Florida to Alaska there are parts of the United States that are today found together in one place, but their geological histories are quite different from one another. We give them the name *terranes*. Florida, all of it, may well have come from Africa, and several places around San Francisco came to that location from elsewhere: Marin Headlands at the north end of the Golden Gate Bridge was identified by its fossils as coming from the bottom of the Pacific far from land. Similar stories describe the history of Alcatraz and Angel Islands.

These are but a few of the many terranes found across the nation. Parts of Nevada, which is quite some distance inland, came from Asia. All along the west coast, from Baja California to Alaska, terranes are everywhere. Two hundred separate ones have been discovered, some large and some quite small, some located as far inland as Utah and Colorado. Most of these terranes arrived between 100 and 200 million years ago, and as a result of their arrival, the continent was extended westward by 500 kilometers. The sources of these new lands are unknown. Some came from ocean floors, some from islands, and some from other continents.

It seems that terranes have been accreted to continents from earliest times, from as far back as 3 billion years ago. They are found today in all continents and in all parts of continents. Some have even suggested that continents are really patchwork rock quilts. The world's most dramatic example of a terrane happened in Asia about 50 million years ago when India, then an island, joined the continent. The event sent repercussions across the world. Ocean crust of the Pacific, which had been moving westward, changed direction and from that time onward spread northwestward.

These terranes now provide valuable clues about rocks that are almost as old as the earth. In fact, they are the only rocks of that age available. Others disappeared in the billions of years during which they were subjected to cycles of erosion and subduction. Using the same magnetic tools that enabled geologists to identify the magnetic strips on both sides of the midocean ridges, as well as their ages from radiocarbon dating, the magnetic orientations and ages of the terranes can be determined. Fossils of plants and animals from the past provide additional data.

TECTONIC PLATES

The huge plates I described as barges glide at differential rates over the plastic mantle beneath. They are relatively solid slabs of rock, and they lie over the curved surface of the planet like a cap on one's head. By the second half of the 1960s this new powerful concept of global plate tectonics was fully documented and universally accepted throughout the scientific community. There are seven large plates, each covering an area many times the size of the United States, and several smaller ones. Each plate moves horizontally relative to others on the softer rock below the lithosphere. Boundaries where plates meet are the places where most seismic activity takes place, and these boundaries can be quite broad, not the narrow spaces suggested by lines on maps.

On the Atlantic side of the United States the tectonic plate moves more slowly than, say, the Pacific plate at its contact with Central America in the area known as the East Pacific Rise. The Atlantic activity is opposite to that of the Pacific, so a tug of war ensues as the Atlantic part of the North American plate pushes westward against Canada and the United States, while the Juan de Fuca plate spreads and then subducts in an easterly direction (Figure 2.4). The Atlantic is winning this contest, and North America is moving toward Asia at a rate of four centimeters per year. Sometime in the geological future, it will become a part of that continent. One factor remains constant in all of this plate activity—the surface area of the earth. Hence the new oceanic lithosphere forming at the midocean ridges must be balanced by consumption of plate material elsewhere.

We map these great plates on flat paper, but, as I indicated, they are moving on a spherical surface and that means parts of them move at different rates at different latitudes. Thus the Atlantic pressure on North America will be greatest in the south and least in the north. If we follow its movements still farther to the north, we find that it hardly moves at all by the time we reach the North Pole. This can be understood by looking at the spacing of lines of longitude as you move away northward or southward from the equator. The spaces between lines narrow until they are zero at the poles. Moving oceanic crust has nowhere to go! The spherical surface is yet another difficulty for geologists when they seek to trace the history of plate movements over the past millions of years.

Action at subduction zones is complex. Over millions of years sediments from erosion of the land and the accretions from the ocean make the point of sub-

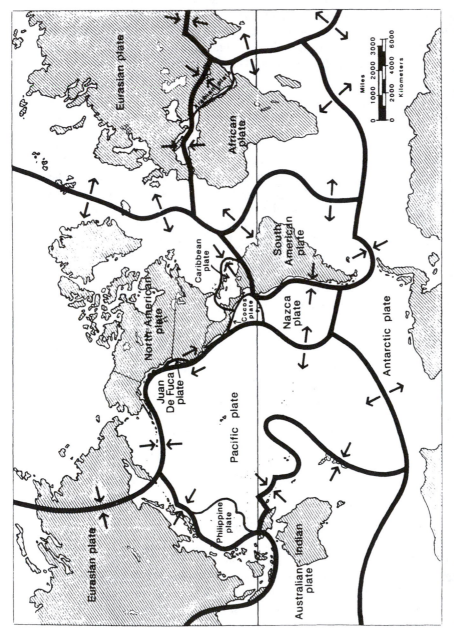

Figure 2.4. The main large tectonic plates around the world. Arrows indicate the directions of movement. The Juan de Fuca plate is moving against the North American plate, gradually subducting beneath it.

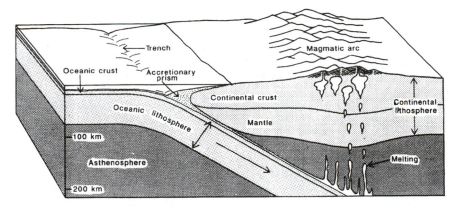

Figure 2.5. Studying the process of subduction closely reveals additional developments: The action begins at the edge of the continental shelf and slope, and a deep trench is created where the oceanic crust (or oceanic lithosphere) goes into the asthenosphere, forcing up either mountains or volcanic eruptions.

duction some distance from the edge of the continent. Thus the slope of the ocean crust is gentle as it approaches the continent and steep at the crunch with the continental mass as it sinks down into the asthenosphere. The accretions, because they are lighter than the ocean crust, often do not form part of the subduction process. Instead, they, along with sediments from land erosion, accumulate near the shore. They fill the deep trench that is formed by the subduction process. Under pressure from the heavier ocean crust as it meets the continental mass, this coastal mass of terranes and sediments may be squeezed upward and pushed eastward from time to time (Figure 2.5). Often in the course of so doing, the intense pressures cause melting in some rocks and metamorphosis in others.

The path of the descending ocean crust is far from smooth, and along its subterranean path, friction and slips trigger earthquakes again and again until the crust reaches a depth of 700 kilometers, where it melts and the quakes cease. In addition to the earthquakes, other things happen on the way down. Various volatiles from water, embedded in the oceanic crust, and sediments from the continental shore dragged down by the descending slab begin to melt soon after 100 kilometers of depth is reached. They then rise into the overlying mantle, which, because it is under less pressure, is melted by the heat from below. Blobs of magma percolate upward and collect in large chambers a few kilometers below the surface.

The magma may sit in these chambers for indefinite periods of time as granitic rocks. Over billions of years it is this process that forms the continents, and because the rocks so formed are buoyant they do not sink back into the mantle. There are granitic rocks formed in this way that date back close to 4 billion years. In places, some of the magma finds a conduit into the surface, and a volcano is the result. In the Pacific Northwest the descent of the oceanic crust takes it over 200 kilometers inland before it reaches 100 kilometers of depth. The magma that reaches the surface here creates the stratovolcanoes that now form the conical mountains of the Cascades, stretching all the way from Mount Garibaldi in southern Canada to Mount Lassen in northern California.

The intensity of earthquakes at subduction sites depends on the age of the oceanic crust. For some parts of Asia, the ocean crust has to travel thousands of kilometers from its spreading center before it encounters a continental mass. That means it must be almost 100 million years old, so cold and dense that it easily slides into the mantle. The opposite is the case with Alaska, the west coasts of North, Central, and South America, and some Asian countries. In these places the subducting oceanic crust is young because the spreading centers are near at hand. The newer and lighter oceanic crust is higher in the water, so a violent collision ensues as it meets the continent. The two plates press against each other, building up greater and greater pressure until something gives way and a powerful earthquake follows.

Dating and Timing

The various explanations and theories about tectonic plate movements require some proof because centuries-old convictions about geology are being discarded for theories that are only a generation old. Convincing evidence of sea floor spreading first came from observations of identical magnetic orientations in successive layers of volcanic rock on either side of midocean ridges. These magnetic anomalies were reversals of magnetism in exactly the same sequences on both sides. In other words, one layer would point north, whereas another farther away from the ridge pointed in the opposite direction. Molten lava loses its magnetism because of the intense heat but regains it as it cools, and the orientation of the new magnetic field, known as *remanent magnetism*, is always the same as that of the earth as a whole at the time of cooling.

However, it is now known that the earth reverses its magnetic field approximately every million years, so remanent magnetism at times will be oriented to the north and at other times to the south magnetic pole. The last time it reversed was 700,000 years ago. By towing sensitive instruments behind a ship sailing at right angles to the midocean ridges, it was found that adjacent strips of ocean, each approximately the same width, kept reversing their polarity. One strip would point to the north magnetic pole, the next to the south, and so on. On the other side of the midocean ridge, exactly the same pattern appeared, a mirror image of the first side, with distances from the ridge the same for each individual magnetic orientation.

All that was needed to complete the evidence for sea floor spreading was some method of dating the rocks. It was essential to show that the strips of rocks with different magnetic orientations were as old as the new theories of plate tectonics demanded. Radioactive decay was the method used. This technique is named after Marie Curie of France who, at the beginning of the twentieth century, described an extraordinary feature of some rocks that are commonly found in volcanic eruption, namely, that they constantly emit energy at a fixed rate and this release of energy is unaffected by temperature or any other changes in the environment. If the amount of energy that is lost by this process of radioactive decay can be captured and measured, then geologists have a very accurate method of calculating the age of rocks.

Geologists were particularly interested in one energy source being radiated from a particular volcanic rock, namely, a gas called argon, because it was possible to measure it. They knew the exact amount of this gas that would be created in a given time. While volcanic rocks are still molten, that is, while they are erupting at the midocean ridges and for some time afterward, the argon gas that is being radiated escapes; but as soon as the rock cools and solidifies, this gas is trapped inside. Over time the amount of gas increases at a constant rate, and as they measured it at widely separated volcanic deposits on the ocean floor, the age of each was calculated.

Metamorphic rocks have to be dated differently because the intense heat associated with their transformation resets, as it were, the radiometric clock, so it has to make a fresh beginning, just as happened when molten rock reached the surface and cooled. The captured gas from these rocks tells us only when they were metamorphosed, not their actual age. Measuring sedimentary rocks is even more difficult. Only rarely can they be dated by radiometric means.

These techniques proved to be particularly valuable when analyzing complex phenomena such as those found in the Hawaiian-Emperor Chain. These same techniques of dating enable us to reconstruct "Pangea." Wegener may have been inaccurate in his representation of an early land mass that later became the world we know today because he had no way of calculating the time frame involved. We can now achieve accurate maps of "Pangeas" at different times. Over long periods of geological time, this map or one like it can recur again and again in the course of earth's billions of years of existence.

The speed at which tectonic plates move today is a very different kind of scale, but again technological developments of the past few decades enable us to measure these movements very accurately. The Global Positioning System, GPS for short, is a set of 21 military satellites that orbit the earth, continually sending out radio signals. A receiver is placed at the location whose rate of movement we want to measure, and a minimum of 3 satellites are used to measure the exact distance from the given location to each. By repeating the procedure over time, the rate of movement is known. At the present time an accuracy within 2 centimeters of error is obtainable, enough to confirm the reliability of the estimated average rate of plate movement, 15 centimeters annually. This seems a very small amount, but in geological time, it is very big, about 150 kilometers in a million years.

The rate of sea floor spreading determines the shape of the midocean ridge. In the Atlantic, where spreading rates are relatively low, the ridge is narrow and the rift valley in the middle is flanked by rugged high ground. In the eastern Pacific, on the East Pacific Rise, the spreading rate is much higher, and so the shape of the ridge, or the rise, is broad with fairly low relief on either side. There are several places where plate movements had a major impact on life in the United States, especially on the west coast where ocean plates collide with the North American one. Juan de Fuca is a good example of these. It is a tectonic plate that had devastating effects on Washington, Oregon, and Alaska through subduction earthquakes and on Hawaii via tsunamis.

Juan de Fuca Plate

From Cape Mendocino in northern California to a point south of the Alaska Panhandle, a relatively small plate, the Juan de Fuca, interacts with both the Pacific plate to its west and the North American plate on the east. Were it not for the expansion taking place in the Atlantic, pushing the North American plate westward at a rate of one centimeter per year, it would be a much bigger plate. At one time it was close to the center of the Pacific Ocean. At its point of contact with the Pacific plate it is a spreading ridge, and as it subducts under the North American plate at a rate of four centimeters per year, it causes an ongoing series of earthquakes.

Subduction earthquakes can be more powerful than any others. The intense pressure of sea floor crust pushing downward against a bigger and more resistant continent can lead at times to an impasse where both plates are stuck and no movement is possible until sufficient pressure builds up to force a slip. The ensuing earthquake often measures 8 or 9 on the Richter Scale, and there have been several of this size along the west coasts of North and South America in the course of the twentieth century. In addition to their usual effects in causing earthquakes, these slips change adjacent land areas in another way, forcing it steadily downward initially over long periods of time where subduction begins and making places farther inland rise. As soon as a slip occurs, the reverse action begins to take effect: Depressed land rises and elevated territory drops down.

The Juan de Fuca plate is particularly interesting from other points of view. In early phases of plate tectonic research, this was the first location to confirm the reality of magnetic anomalies. In the late 1950s, about 10 years before the scientific community finally accepted the now-familiar story of sea floor spreading and moving continents, a coherent pattern of magnetic anomalies was mapped at the Juan de Fuca plate. They appeared like zebras, strips of volcanic rock with alternating magnetic orientation. At first this anomaly was treated as a novelty. Then identical patterns were found on both sides of the Mid-Atlantic Ridge. This was the first set of data to prove the existence of sea floor spreading. The Juan de Fuca plate is still one of the most intensively studied spreading centers anywhere in the world because it contains the full range of features that are present in all of the rest of the world's midocean ridges.

This tectonic plate is now a small version of the bigger original, but it is a vitally important element in the environment of the Pacific Northwest. It is about 500 kilometers long at the ridge and is broken up into six segments ranging in length from 50 to 150 kilometers, a common feature of medium-fast spreading ridges. The most southerly of the six is a little over 2 kilometers wide at its crest, which reaches upward to approximately 2,000 meters below the ocean surface. The rift valley within extends downward about 90 meters. In the center of this valley is a narrow cleft, about 60 meters wide and 20 meters deep, marking the boundary of the Pacific and Juan de Fuca plates. The valley floor is covered with recently erupted volcanic rock.

Residents of Oregon and Washington are more concerned about the subduction activities of the Juan de Fuca plate than its spreading function. They know that someday a massive earthquake will occur along the line of contact with the North American plate. Furthermore, features of the subduction process make this one different from all the others along the western shores of North and South America. The distance that volcanic material has to travel from the spreading ridge to the subduction zone is shorter than any other location on the Pacific coasts of North and South America, so the ocean crust will be relatively thin and warm. This means that the volcanic material will sink at a more gentle slope, and there will be no deep trench or deep earthquakes as is found at other similar sites. Unfortunately for the residents of Washington and Oregon, the lack of a deep trench also means very powerful earthquakes.

Plate History

Working back from all we know about plate behavior and speed, it is possible to trace a good deal of the status of continents and oceans in the past. Along with these discoveries of the last 40 or 50 years, we have data going back much further, thousands of reports by geologists covering all parts of the globe. To begin, we know that the time involved in the change from a Pangea to what we see today is about 200 million years, so it is reasonable to expect that it would take just as long to change back to a single continent configuration. It is unlikely, given the many variables at work, that a second Pangea would be an exact duplicate of the previous one, but the complete cycle of 400 million years is still quite a short period of time, about one tenth of the earth's age, so there should be many Pangeas.

A computer is the instrument of choice for charting the past. The data we now have, such as plate speeds on different oceans, is fed into a program. The outcomes are then calculated somewhat like the experience of running a film backward; we know the end of the film story, but we want to see where the action was at an earlier point in time. Many years ago, at a meeting of the British Royal Society, a computer program was presented in which all the continents were brought together to demonstrate the likelihood of a fit. The result impressed everyone and proved to be a very convincing argument in favor of continental migrations.

About 600 million years ago, according to the best computer estimates available, North America had the equator running north and south through the middle of the continent. What is now Chicago must have been a hot equatorial rain-forest. Siberia lay close by but farther south and Baltica still further to the south but still in a temperate zone. Other continents were much further away to the south from these three, locked together in a single land mass, covered with ice and glaciers. Deep in the soil of Africa's Sahara Desert the evidence of that cold past can still be found.

Some 175 million years later, 425 million years ago, North America and Baltic formed one land mass, and both were on the equator. Siberia had moved northward by this time and was in a temperate zone there. Other continents were still south of North America and Baltic but closer. By 255 million years ago all of the continents were closing in on one another as they approached Pangea. North America had moved northward and looked a bit like today's continent except that the east coast tipped southward close to the equator. As the land that would one day be the United States collided with Europe, its eastern sea-board was heaved up into the future Appalachians.

As the oceans closed in and Pangea became a reality, Europe and the United States found themselves close to the equator. There they were covered with extensive tropical forests long enough to deposit the rich organic matter that became Europe's and North America's coal seams many millions of years later. Here, too, at this time, the dinosaurs roamed, able to move easily from one continent to another. India at this period was far from its present location. It was tightly wedged between Africa and Australia. It would be close to 45 million years ago before it reached Asia, where its energy helped to push up the Himalayas as it met the impasse of the Eurasian land mass.

Pangea did not remain long as a single land mass. In geological terms, it was brief, probably no more than 30 or 40 million years. By 175 million years ago oceans had reappeared. By 40 million years ago the world map looked pretty much as it does today. In North America the Pacific plate, which had been steadily approaching the west coast of the United States, reached the southern shores of the future California, and the long story of the San Andreas Fault was set in motion. From here on, there would be a transform movement of the two plates, the Pacific sliding northward and the North American sliding in a south-western direction.

EARTHQUAKES

From the time that Christopher Columbus reached the shores of the Americas, more than 300 million people worldwide have died from earthquakes, and many millions more have lost their food resources or livelihood. Records go back much longer in all civilizations, recording a sad litany of destruction but with little recognition of the causes of these tragedies. It was not until the twentieth century that there was an understanding of the causal link between rock slip, or faults in rocks, and earthquakes. The breakthrough in our understanding came after the San Francisco earthquake of 1906 when the cause could not be attributed to any force other than a geological fault.

The hallmark of plate movements worldwide is the earthquake, and no other aspect of their activities is more feared. While earthquakes occur in other parts of these huge tectonic plates and in high mountain areas, the vast majority originate around the Pacific Rim, where pairs of plates meet. For this reason the name frequently given to this area is "The Ring of Fire." Southern Europe and Asia are other areas of intense earthquake activity. Each year worldwide there are about 20,000 shallow earthquakes of magnitude 2.5 or more on the Richter Scale. These quakes can be as shallow as 1 or 2 kilometers and as deep as 700 kilometers.

The depths of earthquakes around the edges of plates tell us quite a lot about their origins. Most are no deeper than 100 kilometers because it is within that zone that rocks are brittle, and this is the main environment for quake action. Depth also reveals a lot about the thickness and composition of the relevant plates. The ocean floor bends downward at subduction zones, and in the act of bending, faults that generate quakes are created in the upper part of the plate. These earthquakes are shallow and generally of low magnitude. Quakes continue down into the mantle as far as 700 kilometers. Evidently, ocean crust remains sufficiently brittle to that depth. Thereafter, rock is softened or melted, so that it can no longer rebound in an earthquake. Ninety percent of the seismic energy released by earthquakes worldwide comes from subduction zones.

The first question that is always asked about earthquakes is their size. In 1935 Charles Richter from the California Institute of Technology developed a measure of earthquake magnitude based on the size of the waves detected by a seismograph. He had in mind only the needs of California, but subsequently his scale came into use throughout the world. He used a logarithmic scale in order to cope easily with the huge range of magnitudes among earthquakes: Each number on his scale measured 10 times the strength of the previous number; thus, a 9 magnitude quake would be 10 times as powerful as an 8. All measurements were to be taken at 100 kilometers from the earthquakes' epicenters. In later years, various modifications of Richter's original scale came into service, along with variations on a new one, the Mercalli, which measures not the power of a quake at the source but the damage done at the receiving end.

Eastward from the Pacific coast, all across the United States, there are numerous places that experience earthquakes (Figure 2.6). Stresses caused by the westward movement of the North American plate under pressure from the Mid-Atlantic Ridge and against resistance from the Juan de Fuca plate give rise to numerous faults and thus to opportunities for earthquakes. Historically the greatest concentrations have been in three areas—parts of the Cordillera, the central section of the Mississippi River, and the Appalachians. The continents move as part of the general plate activity but to a lesser degree because of their greater mass.

Additionally, it must be noted that the crust of continents is far from being

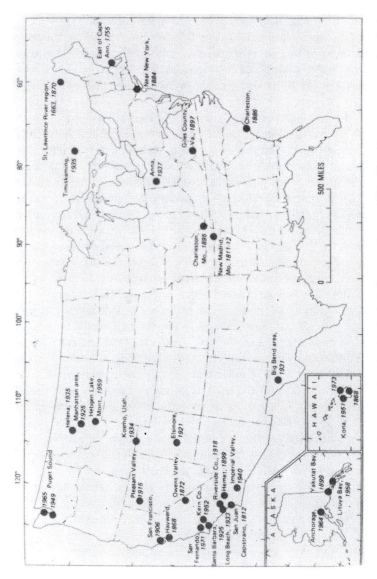

Figure 2.6. Notable historic earthquakes in the United States prior to 1980. The biggest concentration is in California, but significant, major eruptions have occurred from time to time both in other parts of the Cordillera and in the eastern half of the coterminous United States.

an unmovable mass, as was once thought. Wherever there are mountains over 2,000 meters high, it is very common to find numerous earthquakes within the continental plate. There is an inherent instability in all land masses, especially in mountain areas, and so there is always the possibility of movement and the creation of fault lines through which molten lava and gases can escape. If we go back to the idea of Pangea, it is easy to see how a fault line could be the trigger for a cycle of ocean floor spreading. A new ocean would gradually appear as water flows in to cover the newly created sea floor, and a process is launched similar to what we have observed taking place at the present time.

Alaska, 1964

We can gauge the destructive power of an earthquake of magnitude 8 or 9 on the Richter Scale by examining the one that struck Alaska in 1964. Fortunately, the population density of Alaska's southern coast is much less than that of Washington and Oregon. The cause of the Alaskan earthquake, often referred to as the Good Friday quake because it hit on that day, was similar to a 1700 one farther south, which we will study when we come to the Cordilleran Orogen. It was a subduction earthquake caused by the Pacific plate moving northwestward and tipping down into the Aleutian Trench. Numerous landslides occurred, high-rise buildings and bridges collapsed, and tidal waves as high as 50 meters swept over coastal communities, destroying everything in their paths.

The main quake lasted about three minutes and damaged an area of 130,000 square kilometers. More than 130 people lost their lives. Portage became a ghost town because the land subsided 3 meters. Elsewhere, around Cordova, the land rose, in places by as much as 10 meters. The sea floor was raised by an average amount of 3 meters over an area 250 kilometers wide and 800 kilometers long, past Kenai Peninsula and Kodiak Island. Tsunamis created most of the damage in the coastal areas.

They also raced across the Pacific to Japan, Hawaii, even Chile, at speeds of more than 600 kilometers per hour. Sea inlets were flooded all along the coast as far as Washington State. The violent ground shakings that accompanied the earthquake caused slides and avalanches on nearby mountains. Some avalanches were recorded as far away as 240 kilometers. A rock slide more than 200 kilometers away killed a man on Kayak Island. Submarine landslides added to the destruction and appeared as boiling, muddy water that rose hundreds of kilometers above the level of the sea.

San Andreas Fault

From the southern end of the Juan de Fuca plate at Cape Mendocino, there is another type of earthquake-prone region, quite different from the Pacific Northwest. Here the North American plate interacts with the Pacific plate by sliding sideways against it. The San Andreas Fault is the line of contact between these two plates; periodically, as the North American plate slips toward the Northwest, the process is catastrophic because the fault line affects so much of the state. It does not move smoothly.

The San Francisco earthquake of 1906 lives in history as the most dramatic and destructive of the many quakes that have occurred on the San Andreas Fault in the course of the twentieth century. Many buildings were destroyed, especially on the landfill areas near the waterfront. Over 50 fires were ignited all over the city, and they raged for three days, causing far more destruction to buildings than the direct action of the quake. Almost 500 people were killed, and the total cost was $7 billion in terms of current dollar values. The earthquake was felt from central Oregon to Los Angeles, and many towns and communities along the fault line suffered damage.

The two massive plates ground past each other for less than a minute in the early morning of 18 April 1906, with the Pacific plate lurching northwestward by as much as six meters. Strains that had been building for a hundred years were suddenly released. Earthquakes, especially those associated with the San Andreas Fault, are the most dramatic illustrations of the behavior of tectonic plates. Most volcanic eruptions as well as distributions of minerals and fossil fuels are also demonstrations of plate motions. The concept of plate tectonics is indeed the key to many of the earth's secrets. It is a conceptual revolution as profound for the earth sciences as evolution was for the biological sciences.

Fault lines are of different kinds, depending on the earth movements involved, whether they are horizontal or vertical, and in which directions they move. The San Andreas is known as a strike-slip fault, that is to say, it is a slip in a horizontal direction along the line of the fault. This particular fault, since it involves two massive plates, is a long-standing one, and every year it moves several centimeters. When it encounters resistance, pressure builds up to the point where there is enough power to either skip or remove the obstruction. It is at these moments that earthquakes occur, and California is never sure when the next one will come, either from the main fault line or from one of its branch faults.

There are different kinds of waves that emanate from an earthquake. The first is a "P" wave, a push and pull motion just like a sound wave, and it is the one that always arrives first. That's why it is called P, the primary one. The "S" wave, the secondary, arrives a bit later because it is a different kind of wave. It moves at right angles to the direction of the P wave, shaking the rocks around it either horizontally or vertically. It is the more dangerous wave, especially if the place at the surface is on soft ground. S waves expand as they reach the surface and are freed from the limitations of underground pressures, so their destructive strength is directly related to the hardness of the surface rock they finally encounter. There are other waves connected with earthquakes: Rayleigh and Love are two of them, and they act horizontally, along the surface of the ground rather than below (Figure 2.7).

In October 1989 a strength 7 earthquake struck a branch fault of the San

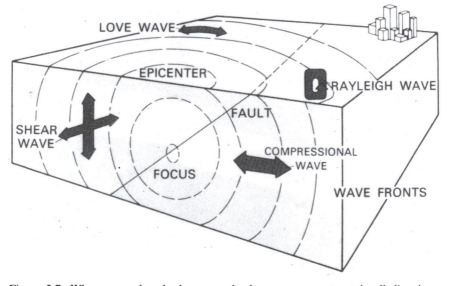

Figure 2.7. When an earthquake happens, shock waves are sent out in all directions. Compressional and shear waves cause high-frequency vibrations, making low buildings vibrate. Rayleigh and Love waves are low-frequency and last longer, effective in making high buildings vibrate.

Andreas near the city of Santa Cruz in the southern Santa Cruz Mountains. A series of thousands of destructive landslides was triggered all along a stretch of coast and central valley from north of San Francisco to points 75 kilometers south of Santa Cruz. Two hundred homes were destroyed along with numerous roads and other buildings. Direct damage was estimated at $30 million. Many of the slopes that failed could have been identified as susceptible to landslides using existing techniques.

Extensive studies were made in the wake of this quake to ensure adequate preparation for any future similar event. For the most part, places that were damaged by landslides were checked out, and where necessary, as in some coastal locations, homes were removed or other remedial action taken. Landslide dams, of which there were several, posed a different kind of problem. Water is stored behind the slide in these places, and for a time, they seem to be secure reservoirs. Later, perhaps as much as a year, there is dam failure and destruction.

One particularly weak area was the Marina District of San Francisco. Despite the experience of 1906 when this part suffered severe damage, city officials went on to fill the area with sand and rubble from the quake, in order to use it as a site for the 1912 Panama-Pacific International Exhibition. In later years it became a very popular section of the city. The lurking danger, which was ignored, was liquefaction in the event of an earthquake. When Loma Prieta struck, the Marina area immediately sank about 12 centimeters. This was followed by

Figure 2.8. Collapsed building in the Marina District of San Francisco due to the 1989 Loma Prieta earthquake. This part of the city is on a former lagoon that was reclaimed for residential settlement. During the quake, ground shaking led to liquefaction, and many homes were destroyed. The same thing happened in 1906, but the experiences of that time seem to have been forgotten by 1989.

widespread liquefaction as water-saturated sand turned into liquid. Buildings shifted off their foundations, and many collapsed (Figure 2.8).

Earthquake Prediction

Earthquake prediction is an inexact science. No one can claim certainty about the future, yet scientists are always at work seeking to gain as much predictability as possible. Occasionally a successful forecast occurs, and then efforts are redoubled to capitalize on the event. In China, in 1969 on a particular morning, zookeepers noticed unusual animal behavior: Swans avoided water, pandas screamed, and snakes refused to go into their holes. About noon on that same day a 7.4 magnitude earthquake struck the city. Ever since then, scientists take careful note of any relevant animal behavior.

Most research on prediction remains focused on the characteristics of rocks and terrain in high-risk areas. Tilting of the ground, changes in elevation, even rising water levels in wells can all be indicators of stresses deeper down. Appearance of small cracks may be the surface manifestation of a fault hundreds of kilometers below ground. In those places that were hit by very powerful earthquakes in the past, it is possible to predict the frequency of recurrence by examining soil and rock layers below ground and calculating the likelihood of another quake. This is a very rough method of prediction, making it possible to say no more than that an earthquake of such and such magnitude will recur sometime within the next so many years.

Although probability seems at first glance to be a very weak method of prediction, it is turning out to be the best of all. When dealing with a very large number of variables in a situation where most of the variables operate independently of one another, the ordinary predictive methods of science do not apply. Scientific prediction requires stability in several variables so that the behavior of one can be evaluated. Weather forecasters were among the first to recognize the value of probability prediction, but the idea did not originate with them. It came from scientists working with very small things, like cells or atoms, where all the common laws of physics give way to random behaviors. Only probability is predictable in those domains.

Meteorologists trying to cope with increases in the frequency and power of both hurricanes and tornadoes decided to settle for probability methods in giving storm warnings. It is now the same in geology. The basis of the method is the record of past events. If there are extensive records of, say, earthquakes in a given region, including magnitudes and dates, it can be said with a stated degree of probability that an earthquake of a certain size will hit within a given time period. This approach has become standard practice over the past 20 or 30 years with the Global Seismograph Network (GSN), which was set up in the early 1960s. In 1990 it rated as high the probability of a magnitude 6 quake striking the eastern United States before 2010.

VOLCANOES

On a worldwide basis most of today's active volcanoes lie close to the margins of the big tectonic plates, and their locations are very often closely linked with the sites of earthquakes. The greatest amount of eruptive activity is of course along the midocean ridges. There, lava is continually emerging from deep within the earth through the rifts in the ridges created by forces pulling the plates apart. Hot magma is lighter than the colder overlying rocks, so it rises by buoyancy. While the volume and number of eruptive sites are greatest at these midocean ridges, there is little destructive activity associated with the eruptions. It is at the convergent, not the divergent, zones that we find the dangerous volcanoes.

Where ocean plates subduct beneath continental ones, volcanoes appear on the overlying plate, 100 kilometers or more from the continental edge. They are caused by the partial melting of rocks in the descending oceanic crust with its water-rich sediments. The high water content in these sediments ensures that oceanic crust will melt by the time it has descended 100 kilometers. The great frictional heat generated by the descending plate gives additional buoyancy to this molten rock, allowing it to ascend to the surface. Often its arrival at the surface is accompanied by explosive rock fragments and ash flows. While these subduction zones are the sites of the world's most damaging volcanoes, it needs to be said that a significant number of powerful volcanoes erupt in the interior of the large tectonic plates. Hawaii is a classic example of these intraplate volcanoes.

Hawaiian Hot Spot

In addition to the cycle of magma from the midocean ridges becoming ocean floor, there are hot spots that rise periodically from much deeper parts of the earth's interior. These plumes provide an explanation for the many volcanoes found in the middle of the large plates and also for long lines of volcanoes like those in the Hawaiian-Emperor Chain. This chain, the result of the Hawaiian hot spot, can be dated, volcano by volcano, by the radioactive decay method. Furthermore, each volcano in the chain has its own magnetic orientation, one that is the same as that of the earth at the time it was born.

The volcanoes of Kilauea and Mauna Loa are on the Big Island of Hawaii. They are active because they are the youngest of a line of volcanoes that stretch for thousands of kilometers to the northwest. The particular hot spot that is operating here in the middle of the great Pacific plate seems to be a fairly permanent thing because over time it has been burning a line of volcanoes as the Pacific plate passed over it. These older volcanoes have sunk a bit into the ocean floor as they cooled, but their history can be traced by examining their rocks: Today they represent 70 million years of activity, stretching 6,000 kilometers all the way from the Island of Oahu, a 3-million-year-old eruption, to the volcano Meiji near the Kamchatka Peninsula in Siberia, 73 million years old. The chain has a bend at the point where the Pacific plate is known to have changed direction.

The two tallest volcanic mountains on the chain lie on the Big Island of Hawaii. They are Mauna Kea and Mauna Loa. Each is more than 4,100 meters high, but if their height above the ocean floor is calculated, each exceeds 10,000 meters, that is to say, higher than Mount Everest. These mountains are the volcanic vent of a broad dome being pushed up by the hot spot below, a dome stretching as much as 500 kilometers on either side of the mountain. From their pinpoints at the tops of the mountains, a river of fine molten basalt rises from time to time, and flows over perpendicular walls of old rocks down into the ocean.

Mount St. Helens, 1980

The Cascade Range of western Oregon and Washington has been an active volcanic arc for the past 36 million years as a result of the subduction activities of the Juan de Fuca plate. Evidence for recent eruptions can be seen in the many volcanic peaks from Mt. Baker near the Canadian border to Mt. Shasta in northern California, all of them as yet only slightly affected by erosion. Over the past 40,000 years Mount St. Helens has been the most active of all these peaks.

Mount St. Helens is also the youngest. The cone that partly collapsed in 1980 was only 2,200 years old. Its historical pattern of eruptions consists of four periods of intermittent action, each lasting for centuries or even thousands of years, then followed by many more thousands of years of inactivity before the next cycle began. At the present time it is clear that the mountain is in one of its active phases, and fortunately the events of 1980 were not a total surprise. Few mountains have been studied so carefully. Thus, when a pronounced bulge began to form on the north flank in March, changing shape at the rate of five meters per day, people were moved away from nearby areas.

Early in the morning of 18 May 1980, Mount St. Helens exploded. The earthquake measured 5.1 on the Richter Scale, a strength that would not normally be recorded as catastrophic, but in this case it initiated one of the largest landslides and eruptions ever recorded in the United States. The bulge launched a giant avalanche of rock and mud, and this was followed by a blast of gas and ash that rose 20 kilometers into the air and drifted east across the country. In minutes the mountain's height was reduced from about 2,900 meters to approximately 2,500 and its cone transformed into a gaping crater (Figure 2.9).

Day turned into night in many nearby communities as the ash cloud spread. Over 600 square kilometers of forest were destroyed. Altogether some three cubic kilometers of rock, mud, ice, and ash were moved out of the mountain in one dramatic explosion. Mud flows swept westward down river valleys toward the Columbia River, blocking long sections of the channel for oceangoing ships. Many homes were destroyed, and 57 people lost their lives. In the years that followed, a dome of lava began to build up in the middle of the crater. In time, it will completely fill it.

NATURAL RESOURCES

Minerals

Most of the metallic minerals mined in the western United States—copper, molybdenum, tungsten, gold, silver, lead—formed from magmas above subduction zones less than 100 million years ago. The Pacific plate was sliding rapidly beneath the North American plate at that time, and belts of volcanoes appeared, within each of which granites crystallized. Metals were concentrated in the last of the liquids in the magma chambers. Different combinations of metals concentrated at different levels in the magma chamber, so the depth of erosion that takes place in an igneous complex determines the type of ore deposit found within it today.

Copper occurs near the tops of granite masses that crystallize near the surface after their magmas had risen through much of the overlying crust. Silver tends to be found in the throats of volcanoes, whereas tin forms from granites that

Figure 2.9. Ash from the massive eruption of Mount St. Helens in 1980, one of the largest ever recorded in the United States. The ash was carried vertically to a height of 20 kilometers, and then it circled the globe for two weeks. Horizontally it swept across the country at a height of 10 kilometers.

are created by melting within the crust rather than from melting within the mantle. Some minerals are found in zones of tectonic accretion, that is to say, in rock masses that were ferried by oceanic plates and deposited on continental margins. Nickel and mercury are two minerals often found in these places. Secondary processes can often obscure the influence of plate tectonics. Gold may be present in tiny quantities in subduction complexes but may be found in high concentrations as placer deposits in streams. Uranium ores in Wyoming and the Colorado Plateau may have been formed from volcanic ash blown 800 kilometers inland from west coast volcanic belts.

Oil and Natural Gas

These two resources are found in sedimentary rocks that are by-products of the motions of tectonic plates. Both oil and natural gas are created by the break-down of organic matter, and over time, depending on the temperatures and pressures to which they are subjected, they change into gas and petroleum. These then migrate upward through pores in the rock and accumulate where migration paths are blocked by an impermeable layer of rock. Only a minute portion of the earth's sedimentary rock contains recoverable amounts of either petroleum or natural gas because of the many conditions that must be met. For instance, most petroleum occurs in rocks less than 200 million years old in porous sand-stone or limestone that is shallower than 3,000 meters.

Coal

Like oil and natural gas, coal is created from wood and other organic material that is buried beneath sediments and transformed slowly under the influence of heat and pressure until it becomes a high-carbon residue. Most coal is less than 350 million years old, and the quantities recoverable, on the world scale, have a total energy content that is far in excess of the total energy available from petroleum and natural gas. The Appalachian coal deposits formed about 300 million years ago.

At approximately 200 million years ago the Atlantic Ocean began to open. Oceanic activity developed at the same place that had previously been a point of contact and collision between Africa and North America. In the course of the ocean's formation, a low-lying area bordering the eastern United States was repeatedly flooded by the sea, then drained, and in the periods between these alternating events, coal deposits were created, adding to the quantities formed during Pangea. The western coal deposits of the United States have a different history. They are less that 80 million years old, much younger than those of Appalachia. Some of these younger coals are found in unusually thick layers. One of them is 70 meters thick.

Mitigating Damage

One big question relates to the costs of earthquakes. Since the times of occurrences and the extent of damage cannot be predicted, how can insurance companies provide coverage? What happens if these companies are unable to pay claims and go bankrupt? Perhaps it might be more appropriate to ask, What happens if insurance premiums keep rising, as they have done over the past several years, to the point where most people cannot afford them? In my introduction, I pointed out some of the things that lead to greater probability of damage and costs over the past 30 years to confirm that dangers are increasing disproportionately. Total economic losses in the United States from natural disasters in the 1960s averaged $3 billion annually; in the 1990s they were running at $30 billion per year. Is this due to increased population or poor planning? It is a question that needs to be considered seriously at all levels of society.

Some have suggested that people who deliberately locate in an area known to be dangerous should be expected to bear some of the costs and not leave everything to the insurance companies. National action, in the form of the Federal Emergency Management Agency, has already been taken at the governmental level to minimize loss of life and property and to give immediate assistance in a disaster. State and private agencies have sprung up for the same purpose. Our knowledge base has been improving but is still far short of what is needed. There are large areas of the country where nothing has happened for decades, and perhaps because of that, little research is being undertaken. In California, because of its high profile regarding earthquakes, there is a high level of knowledge about and research into causes of earthquakes, but this pattern needs to be extended to other parts of the country.

One valuable step was taken by the city of Los Angeles following the magnitude 7 earthquake of 1971. It points the way to steps that many other cities and states could take. Unreinforced masonry buildings were the target. It had been known for a long time that these buildings could collapse completely in an earthquake, the walls falling outward, leaving the roofs without support. Occupants get crushed before they have a chance to get out of the building. Laws were passed requiring owners to bring their buildings up to safe levels.

The story does not end there. Opposition to the new building code was swift. Most of the unreinforced buildings were occupied by poor, often immigrant, families, and the prospect of forcing them out of their homes to allow repairs to be done, at a time of severe housing shortage, proved to be too costly politically. Few buildings were renovated. It was a similar story in other cities of southern California. The lesson learned was that mitigation efforts must be done far in advance of and not just as a response to disastrous events.

In the next three chapters, all of the examples of earthquakes and volcanic eruptions that we examined briefly—the Juan de Fuca plate, Alaska in 1964, the Hawaiian hot spot, Mount St. Helens in 1980, and the San Andreas Fault—will be studied in greater detail. They not only represent the common elements in the earthquakes and volcanic eruptions that Alaska, Hawaii, and the coterminous United States have experienced, but they are also representative of the seismic disruptions that are found in other parts of the United States.

REFERENCES FOR FURTHER STUDY

Bolt, B. A. *Earthquakes*. New York: W. H. Freeman, 1993.

Coburn, A., and R.J.S. Spence. *Earthquake Protection*. New York: John Wiley and Sons, 1992.

Dalrymple, G. B. *The Age of the Earth*. Palo Alto, CA: Stanford University Press, 1991.

Hunt, C. B. *Natural Regions of the United States and Canada*. San Francisco: W. H. Freeman, 1974.

King, P. B. *Geological Evolution of North America*. Princeton, NJ: Princeton University Press, 1977.

Lamb, Simon, and David Sington. *Earth Story: The Shaping of Our World*. Princeton, NJ: Princeton University Press, 1998.

Menard, H. W. *Ocean of Truth: A Personal History of Global Tectonics*. Princeton, NJ: Princeton University Press, 1995.

Moores, E. M., ed. *Shaping the Earth: Tectonics of Continents and Oceans*. New York: W. H. Freeman, 1990.

Sieh, Kerry, and Simon LeVay. *The Earth in Turmoil*. New York: W. H. Freeman, 1998.

Thornbury, W. D. *Regional Geomorphology of the United States*. New York: John Wiley and Sons, 1965.

ALASKA

Alaska stands apart from the other 49 states in many ways, especially in relation to the main theme of this book, the impact of geology and associated climatic elements on U.S. environments; it is the volcanic capital of the nation, with more active or potentially active volcanoes than all the other states put together. There are two reasons why the media do not focus as sharply on eruptions here as they do on the coterminous United States. The first is because most eruptions occur in thinly populated or unpopulated areas, along the Aleutian chain of islands. Second, only a small number of people are affected even in the main settled areas; there are only 1 million people in the whole of Alaska.

Curving westward from Anchorage for 1,000 kilometers is the Alaska Peninsula, and beyond it for a further 1,500 kilometers lie the Aleutian Islands. It is on this curve, marking the boundary between Pacific and North American plates, that the majority of volcanoes lie. The islands of the Aleutians are all created by subduction on the part of the Pacific plate. These distances are indicative of the enormous size of the state of Alaska, in area more than twice the size of any one of the other 49. No less impressive is the number of volcanic eruptions over this 2,500 kilometers of territory during the past 200 years, on average, one every year. We will study some of these eruptions.

Because of its northerly location, Alaska is dominated by the problem of permafrost, ground that is frozen to some degree all through the year. No other state has to cope with this particular problem. It is something that affects everything in the day-to-day lives of its inhabitants. Only the fringes along the south and southeast coasts are free from this condition. Even more serious for Alaskans is the threat of global warming because whenever the world's average temperature rises 1 degree, Alaska's rises 1.1 or more because of world patterns of circulation. Already Alaska's average temperatures have risen higher than anywhere in the other 49 states, and the effects are causing alarm.

Finally, there is one more thing that sets Alaska apart from the other states; it is the way in which the Pacific plate meets and interacts with the North American plate. Alaska has the Pacific plate subducting beneath the North American one, just like Washington, and, at the same time, in the same general location, has strike-slip plate contacts somewhat like the San Andreas Fault in California. For this reason, I placed this chapter immediately after the second one in order to emphasize these similarities between Alaskan events and the ones in Washington and California, which are described briefly in Chapter 2.

GEOLOGICAL OVERVIEW

The Pacific coasts of North America and their hinterlands are shaped by interactions between the Pacific and North American plates. The forces at work affect all four states—Alaska, California, Oregon, and Washington—and the mountainous ranges that we see today running all the way from Mexico to Alaska are the end products of these plate interactions. Each state is affected differently, and we have seen examples of the different outcomes in the previous chapter, but the overriding influences are the same. It is almost impossible to separate out any one of the four states and define plate actions that are unique to that one place.

For Alaska, two of the characteristic interactions come together in one place, a rare happening: The western part of the Gulf of Alaska is a subduction zone, the long-term products of which include the Aleutian Islands, and the eastern area is dominated by the thousand-meter-long combined Queen Charlotte and Fairweather Faults, on the sides of which plates are moving in opposite directions at a rate of six centimeters a year. Between the two is a very complex region, as you would expect, and geologists are still trying to make sense of it. Figure 3.1 shows the locations of the two pairs of interactions along with the transition fault region where the uncertainties are focused. The numerous faults between the two zones are further evidence of the enormous complexity of plate movements here.

Nowhere else in the United States have the dramatic effects of subduction been more powerfully demonstrated than in Alaska. While pressures from plate movements occur elsewhere in the coterminous United States, involving major earthquakes on occasion, the extent and in some cases the intensity of Alaska's plate movements in historic times exceed them all. For these reasons, the 1964 earthquake that was briefly introduced in Chapter 1 will be examined in detail. Some of the many powerful volcanic eruptions that occur here will also be given substantial coverage.

Sometime in the future, a quake like the one in 1964 will hit the Pacific Northwest. We will study this prospect in Chapter 5, and Alaska's experience will be a valuable introduction to it. Like the Alaskan, this future one will be

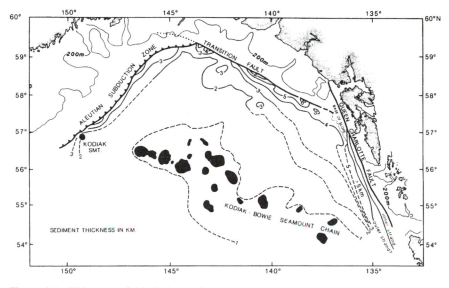

Figure 3.1. This map of Alaska's southern coast shows the two pairs of tectonic plates that are active. At the Aleutian subduction zone the Pacific plate pushes under the North American, and along the Queen Charlotte Fault, the Pacific and North American plates slide past each other like the action at the San Andreas Fault in California.

violent because its subducting rock is young, having come from the nearby Juan de Fuca plate, the one that caused the 1964 quake. Sea floor spreading from the Juan de Fuca is carried by the Pacific plate northwestward the relatively short distance from the spreading ridge to both the Aleutians and the Alaska Peninsula, wreaking havoc as it arrives all along that coast.

CLIMATE

The presence of numerous glaciers near the shores of the Gulf of Alaska westward from Glacier Bay is a good indication of the prevailing climatic conditions: La Perouse, Fairweather, Grand Plateau, Bering, and Malaspina are some of these, a few of them named after the explorers from Russia and Europe who ventured into this area in the nineteenth century. Malaspina Glacier has received a lot of attention because of its size, 1,000 meters deep in places and covering an area of more than 4,000 square kilometers. With the aid of radar it is possible to detect the shape of the land surface at the very bottom of the ice, so Malaspina, because of its massive size, can serve as a model of the ice sheets of the Pleistocene epoch. Thus, year by year, the behavior of this glacier is providing new insights into the ways in which the land was being gradually transformed in the times of the ice ages.

Across the southern part of Alaska, in places like Juneau and Kodiak, there

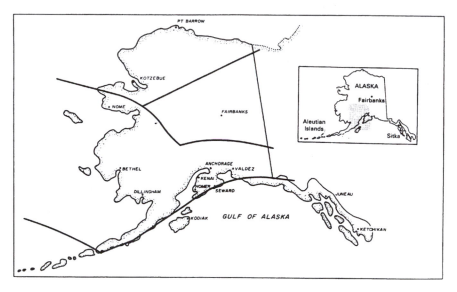

Figure 3.2. The shaded area on the *inset* map marks the area where earthquakes do most damage. Elsewhere, population densities are low. There are four climatic zones in the state represented by Pt. Barrow, Fairbanks, Anchorage, and Juneau. These climates range from year-round permafrost in the far north to mild maritime conditions in Juneau.

is a relatively warm maritime climate similar to that of coastal regions farther south in Washington and Oregon, due to the powerful warming effect of Pacific Ocean currents. Small temperature variations characterize this type of climate, with high humidity and high precipitation. To a lesser extent the ocean influence is felt in the next climatic zone, represented in Figure 3.2 by such cities as Anchorage and Bethel. Here temperature variations are greater, and the annual mean is about minus one degree Celsius, compared with four degrees in the maritime region.

Farther north, around Fairbanks, past the mountain barriers that limit both the amounts of precipitation and the influence of the Pacific Ocean, the climate is true continental, with great diurnal and annual temperature variations and a mean annual temperature of minus seven degrees Celsius. Still farther north, from Kotzebue to Pt. Barrow, the climate is Arctic. There is very little precipitation here, and the mean annual temperature is minus 10 degrees Celsius. Winds are frequently strong.

VOLCANIC ACTIVITY

Earthquake of 1964

Late afternoon on 27 March 1964 it struck, ironically named the Good Friday earthquake because of its proximity to Easter. Figures 3.3 and 3.4 give us a

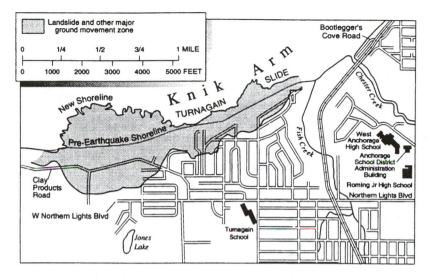

Figure 3.3. In the 1964 earthquake, the bluff along Knik Arm in Anchorage, Alaska, disintegrated as a result of ground shaking. More than 75 homes in the Turnagain Heights were destroyed when the bluff failed. The map shows the change of shoreline caused by the earthquake.

glimpse of the terrible event. It was a subduction earthquake, like all the others that have so often shaken Alaska's Peninsula and the Aleutian chain of islands, but this one was exceptional, the biggest in living memory, with a magnitude of 9 on the Richter Scale. The epicenter was close to the coast in Prince William Sound, about halfway between Portage and Valdez, and from the fishing port of Cordova all the way to the Island of Kodiak, there was one swath of destruction.

Numerous landslides occurred, huge rocks crashed down the mountainsides, high-rise buildings and bridges collapsed, and tidal waves as high as 50 meters swept over coastal communities, carrying away everything in their paths. One small island, Middleton, was uplifted 5 meters. An old ship that had been sunk nearby many years earlier and was normally out of sight even at low tide suddenly was pushed upward and left high and dry above sea level. Fortunately population density here is low, so the death toll was light. It would be a very different story if the quake had happened in San Francisco or Los Angeles.

Figure 3.3 shows the extent of a landslide that overwhelmed Turnagain Heights, a subdivision of 75 homes, many perched on a bluff with an excellent view of the ocean. The bluff disintegrated and tumbled down the slopes into Knit Arm, taking the 75 homes down as it went. All roads and utility lines were rendered unusable. Strange as it may seem, after the earthquake private developers were allowed to rebuild on the slide mass, provided they assumed full responsibility, including liability in case of accidents, for constructing and maintaining new roads and utilities.

Figure 3.4. Scene from Anchorage following the earthquake that devastated southern Alaska in 1964. As a subduction quake, it was an extremely powerful one, measuring near the top of the Richter Scale. This photograph shows a section of central Anchorage that was destroyed due to the settling of unconsolidated sediments.

Anchorage was hit harder than any other city (Figure 3.4). People still remember the way the shocks persisted, each minute feeling like an hour and the dislocation of everything worsening by the second. Underneath each building and everywhere around on the outside, the ground rose and fell like the waves of the sea. Blocks of houses slid about, pavements burst open, and huge fissures opened up in the ground. People clung to lampposts and anything else that had any degree of stability. Broken glass covered the ground. Concrete slabs breaking off buildings killed some who were walking below. Others were killed in their cars as they drove past disintegrating buildings.

At Valdez where a ship was unloading supplies, the dock where it was berthed suddenly collapsed as the whole structure was sucked under, taking two dozen people with it. None of them survived. The shoreline all along the waterfront had also collapsed, and water surged violently backward and forward within the harbor, crushing whatever was left of shore installations. Hours later, to the astonishment of everyone, a huge tsunami washed up on Valdez. Reasons for the time delay will be explained in the next chapter when we study Hawaii, a state that experiences more tsunamis than any other.

The first crest of the tsunami had arrived about 30 minutes after the earthquake, but no one noticed it because it was low tide and the water reached only as far as the high-water mark. Then, about four hours later, wave after wave poured in, at intervals of 30 minutes each, flooding the town of Valdez and wrecking all of its commercial buildings and half of its homes. Fortunately, almost everyone by this time had fled into the surrounding high ground. They stayed there all night in temperatures that dropped below zero, returning to town in the morning to search for lost relatives. Some 32 lives had been lost.

West of the epicenter, at the town of Seward, a similar story to Valdez's unfolded with devastation from both earthquake and tsunami. Seward is an oil port and rail terminus. Again the waterfront structure disappeared beneath the water, taking the installations on the dock down with it. Farther back diesel locomotives weighing more than 100 tons were thrown on their sides. To make matters worse, there were storage tanks of gasoline at the dock that caught fire and added to the confusion. Most of Seward's residential and commercial buildings and all of its industrial areas were obliterated.

The main tsunami generated by the earthquake moved out across the Pacific, first to flood low-lying areas in the town of Port Alberni on Vancouver Island's west coast, then to crash onto beaches in Oregon and California. That was not the end. Once the wave reached deep water, its speed accelerated to more than 600 kilometers an hour. Within six hours it was in Hawaii and after a few more hours, in Japan. The range of the earthquake's influence was evident in other ways. Buildings swayed in Seattle, some ground movement was observed in Texas, and similar though weaker movements were felt in Florida. In Alaska, because of the enormous power of the event, hundreds of aftershocks occurred within a few days. Thousands more followed over the following 12 months.

Liquefaction had caused widespread damage to railroad tracks and bridges to a degree that geologists had never before observed. There were landslides, ground cracks, and warping of the surface almost everywhere, but they were surprised by the degree of twisting in different directions along railroad tracks and the contortions in bridges. Under pressure from ground failure on both sides of streams, as liquefaction allowed these pressures to build up, bridges buckled in their centers. Perhaps liquefaction played a part in the failure of Turnagain Heights, the area shown in Figure 3.3.

In summary, the amount of Alaska and its neighboring coastal waters deformed by the quake was more than 250,000 square kilometers. Some of it had been pushed upward, whereas other locations were dropped down. In places extending as far as 150 kilometers inland, the land was dropped 2 meters, whereas toward the sea, land was raised 2 meters and in some places 10 meters. Along the subduction line a wide stretch of sea floor was heaved up 15 meters, and it was this sudden displacement of water that created the huge tsunami. Furthermore, there were lateral movements of land toward the sea from the west and north, 60,000 square kilometers of it. Anchorage shifted 2 meters, Valdez 10, and Seward 14.

Redoubt Eruption, 1989

In sharp contrast to the violent and surprise character of earthquakes such as the 1964 one, which fortunately are rare events, volcanic eruptions give some warning of their impending action. That is good since there are so many of them in Alaska. Redoubt is a steep-sided peak more than 3,000 meters high, located about 175 kilometers southwest of Anchorage in the rugged Lake Clark National Park and Reserve. Although it is in the mild climatic zone of Alaska's southern coast, its height ensures the presence of 10 glaciers that radiate from the general area of the summit, including an ice-filled crater almost 2 kilometers in diameter. Ice from this crater flows downhill as the Drift glacier for 8 kilometers before becoming Drift River. About 35 kilometers farther south, the Drift River enters Cook Inlet close to an oil-storage terminal housing 2 million barrels of oil.

It is easy to imagine what might happen if red-hot lava or ash erupts from a volcano that is surrounded by ice and snow. Alaskans know this well, and we will see some of the things that can happen as we consider the events of 1989 and 1990. As a precaution against the possibility of flooding if Redoubt did erupt, the oil-storage terminal on Cook Inlet was protected by a 4-meter-high levee on two sides, and to prevent oil spills, each storage tank was surrounded by a 2.5-meter-high terrace. Over the past 10,000 years numerous eruptions occurred in and around Cook Inlet, as evidenced by the buried layers of volcanic ash. At least 30 of these layers are attributed to an equivalent number of eruptions of Redoubt. Research findings of this kind persuaded the federal government to set up the Alaska Volcano Observatory in 1988, similar to the one established in Hawaii at an earlier time, to monitor volcanic unrest.

It was fortuitous that the Observatory was established when it was. It had just completed a series of detailed assessments of hazards when Redoubt exploded into action in mid-December of 1989. The day before the eruption, there was a swarm of earthquakes beneath, giving enough warning to those nearby to take evasive action. A huge cloud of ash out of the summit was the first thing to happen. Tons of debris and gas swept across the neighboring glaciers, leading to massive flows of water and rocks down Drift River Valley. In the weeks and months that followed, there were a series of eruptions of different magnitudes and with different types of materials ejected. Before action ended in August 1990, major damage had been done to the oil-storage terminal and other places. Overall costs rose to $100 million, making it the second-most costly eruption in the history of the United States, exceeded only by Mount St. Helens in 1980.

In the first day there were four explosive episodes, each producing columns of tephra, some of which rose as high as 12,000 meters. *Tephra* is a term that refers to anything airborne from an eruption and includes ash-size fragments as well as the larger rock fragments derived from inside the volcano. The tephra was carried by wind all over the southern coastal area, leaving layers of gritty ash in its wake. At the same time, repeated flows of water and debris roared down Drift River as ice and snow around the summit melted. All of this swept past the storage terminal, and while the terminal was not washed out, conditions were so dangerous that all of the maintenance crew were evacuated. By the end of the day the volcano's vent had been cleared of old rocks, and magma with pumice was escaping to the surface.

Over the following week or two a lava dome took shape within the volcano as magma continued to erupt. This lava was semiliquid, a common feature of stratovolcanoes. Redoubt is a stratovolcano just like those of the Cascade Range in the Pacific Northwest, and the dome that took shape in Redoubt was similar to the one that appeared in the crater of Mount St. Helens in the months following the initial eruption. By the first of January the new lava dome had grown to a point where it was blocking the escape of the underlying magma. The Alaska Volcano Observatory warned that an explosion was imminent, one that would collapse the dome. The next day it happened. About 80 percent of the dome was blown away in a double explosion described by one onlooker as an orange flame shooting straight upward from the volcano's summit. New masses of debris inundated the Drift River, hundreds of new gullies and crevasses appeared on the volcano's sides, and the valley floors of some streams were eroded down to bedrock by the speed and roughness of the debris material.

Because of Redoubt's behavior, all domestic and international airline traffic in and around Anchorage was halted, and this meant major changes to international routes since Alaska is an important stopover for flights between the United States and Asia. Furthermore, the flights had to be canceled or rerouted at short notice around Christmas, a time of heavy traffic. It was estimated that airlines lost close to $3 million on account of the eruptions of Redoubt. An even bigger question arises in relation to the risk factor associated with airlines flying through an area subject to volcanic eruptions.

Volcanic Eruptions and Aircraft

Between December 1989 and January 1990, before the dangers from volcanic ash were fully recognized, four commercial jet aircraft encountered ash downwind from Redoubt. The abrasive quality of the ash damaged the cockpit windshields and the leading edges of the wings so seriously that the windshields had to be replaced and the wings polished. All the engines required a thorough inspection. The reason for all this is speed. At a jet's cruising speed, even a speck of dust can make a dent in the metal frame of an aircraft. An earth satellite, traveling so much faster than a jet, can be put out of action completely by particles that would not be noticed if they had struck the windshield of a car.

The most serious incident happened on 15 December 1989 when a Boeing 747 en route from Amsterdam, carrying 231 passengers and a crew of 13, began its descent into Anchorage. Another 747 had followed the same descent path only 20 minutes earlier and had landed safely. The plane from Amsterdam, however, ran into a cloud of ash 240 kilometers downwind from Redoubt at an altitude of 7,500 meters. The volcano had erupted about 90 minutes earlier. As the pilot attempted to climb out of the ash, some particles that were melted by the heat of the engines began to solidify, forming a glassy coating on the turbine blades, thus restricting air intake. All four engines shut down.

For the next 8 minutes the plane glided steeply, losing 4,000 meters of altitude before the pilot succeeded in restarting the engines with less than 2,000 meters of clearance between the plane and the ground. The cold air of Alaska and the repeated attempts to restart the engines resulted in a partial breakup of the glass coating. The plane landed safely in Anchorage about 40 minutes after the incident began. All four engines and the electrical circuits had to be replaced and all the fine ash removed. The total cost of these repairs was $80 million. This near-tragic encounter, augmented by the news that as many as 23 planes had similarly encountered volcanic ash in different locations over the previous two decades, heightened awareness of the danger.

New efforts were launched by the Alaska Volcano Observatory to make sure, as far as was humanly possible, that an event of this kind would not happen again. There are monitoring stations all along the volcanic arc, with special additional sites close to the main airports where communications are maintained continuously with incoming planes. Exact data are provided to pilots whenever an eruption takes place. Since it is now possible, through observation of the frequency and strength of earthquakes immediately prior to an eruption, to predict to within a few hours when the eruption will occur, pilots will not again be taken by surprise. Warnings will reach them long before they begin their descent.

Augustine Eruption, 1976

On a small uninhabited island in Cook Inlet with the same name as the volcano, a name given to it by Captain Cook in 1788, about 130 kilometers south of Redoubt, stands Augustine volcano. It is, like Redoubt, a stratovolcano, so it has a symmetrical, cone-shaped appearance, and over the past two centuries, it has erupted more frequently than any other in this southern part of Alaska, the area where most people live. It was active in 1812, 1883, 1902, 1935, 1963, 1971, and 1976. It was also active later, in 1986, but the 1976 event may be a better one to study, because it is typical of the type of volcanism experienced in Cook Inlet. Like Redoubt, advance warnings of impending action come in the form of swarms of earthquakes deep down within the mountain. Even in the days before the Alaska Volcano Observatory, monitors picked up these warnings in Augustine.

The 1976 eruption began late in January, and over the succeeding three days, six major explosive eruptions took place. Ash was repeatedly blasted skyward to heights of 13,000 meters, then fell back to cover an area of more than 200,000 square kilometers. In between the explosive episodes, ash and gas avalanches swept down gullies in the mountain's sides, often at speeds of 150 kilometers per hour, to end up in Cook Inlet. Internal temperatures in these flows reached 700 degrees Celsius.

For about 10 days everything seemed to be at rest, but in the first week of February, it started again. There was another series of explosions that continued for 10 days, and just as happened with Redoubt, the old dome that had been built up after the eruptions of 1964 and 1971 was finally destroyed in this second phase. A new dome began to grow soon after the last explosion and within a short time had reached a height of 250 meters. This speedy rate of growth suggested that a large magma pool must lie inside the mountain at a relatively shallow depth. If so, this would explain the frequency of Augustine's eruptions.

The destructive effects of the widely distributed ash were felt in many places: A high school cross-country ski team found that the ash was stripping the wax from their skis, and those who had contact lenses found their eyes becoming irritated as tiny flecks of dust got behind the lenses; electric utility officials noticed that their natural gas–powered turbines increased their efficiency as the ash rubbed a layer of corrosion off the turbine blades. Useful though the increased efficiency was, they feared that further action by ash could destroy the blades, so they stopped the turbines until the ash could be cleaned off. Analysis of the gas showed that it consisted of tiny fragments of glass and such minerals as magnetite and plagioclase. It also had small percentages of sulfur and phosphorous in similar proportions to those found in the environs of Mount Katmai after its 1912 eruption.

Mount Katmai Eruption, 1912

Augustine is a good example of the frequency of volcanic eruptions in and around Cook Inlet, but Mount Katmai, a, 2000-meter peak, is an example of quite a different kind: It was the world's largest twentieth-century eruption and certainly Alaska's worst ever in historic times. Mount Katmai and its sister mountain, Novarupta, are about 150 kilometers southwest of Augustine and 500 kilometers southwest of Anchorage, within Katmai National Park and Preserve. When the eruption occurred, after centuries of silence from both of these mountains, 24 cubic kilometers of ash and small particles were blown into the sky, giving rise a few years later to a new name for the area, "Valley of ten thousand smokes."

Five days before the eruption, small earthquakes began at Katmai Village, about 30 kilometers from the volcano. Villagers from Katmai and other nearby communities moved away to a safer location. Perhaps they knew from previous experience that conditions would be intolerable at 30 meters from the volcano. The eruptive activity began in June of 1912 and continued intermittently for two days. The violence of the earth tremors that accompanied the eruptions was felt as far away as Seattle. Ash was so heavy that darkness fell at midday over an area extending 200 kilometers east of Katmai. Around Katmai itself for about 20 kilometers in all directions the ash was 2 meters deep.

A ship moored in Kodiak harbor at the time of the eruption gave this account of events. Twelve centimeters of ash has now fallen everywhere, choking all wells and streams on shore. Visibility dropped to 16 meters, and after about two more hours, at a time when the sun would have been shining brightly, pitch darkness set in and continued into the morning of the next day. Decks, masts, and lifeboats were all covered with a fine, yellowish-colored dust. Avalanches of ashes could be heard sliding down the neighboring hills and sending out clouds of suffocating dust. The crew kept working continually with shovels and hoses to try to rid the ship of ash. The dust was so thick that a lantern could not be seen at arm's length. The amount of ash that had fallen on Kodiak was later calculated at 30 centimeters.

Like other gigantic events of this kind, where ash and other materials are flung high into the upper atmosphere, the effects of Katmai's eruptions were felt far and near. Locally 80,000 square kilometers had a covering of three centimeters of ash. Smaller amounts fell on Juneau, more than a thousand kilometers to the southeast, and still less on Seattle, 2,400 kilometers to the south. Some ash particles reached far enough into the atmosphere that they were caught in global circulation patterns and provided spectacular red sunsets for months.

GLACIERS AND PERMAFROST

Glaciers are found in many places where there is no permafrost, but they are most numerous in very cold climates, that is to say, in northern latitudes or in high mountain areas. Alaska has lots of them (Figure 3.5). All that is needed for a glacier to form and then become active is the existence of a line at a certain elevation where more snow and ice are added in the course of the year than are lost. This line, which we could call the melt point, is not fixed. It moves up and down with the seasons, and it also moves over time as changes occur in the climate. As more and more snow accumulates above the melt line, the glacier is pushed downhill past the point where it would ordinarily melt, the distance beyond this point depending on the thickness of the ice and the prevailing temperature.

The glacier is sensitive to pressure as well as temperature so that, where the ice is thick enough, pressure at the base causes melting, and a stream forms. This stream of water can sometimes be a roaring river emerging from beneath the glacier at the melt point. The erosive power of such a stream is enormous because it is cascading down the steep slopes of a mountain and because water's carrying and gouging energy increases dramatically as its speed increases. When accompanied by the downward movement of ice, which is an even greater force, the valley that was formerly V-shaped is now U-shaped.

Permafrost is defined as soil or rock with temperatures that remain below zero degrees Celsius for two or more years. It is widespread in Alaska, affecting 80 percent of the state. On the southern fringes permafrost appears as thin, isolated clusters. As one goes farther north, its depth increases until it reaches down as far as 500 meters. At the same time the area affected becomes continuous. Permafrost comes in many forms. It may not be frozen at all if it is dry, that

Figure 3.5. An Arctic glacier near the southern Alaska-Yukon border.

is, having neither ice nor water. It may contain unfrozen water or a mixture of unfrozen water and ice, or it may have ice only.

A home built on permafrost cannot be constructed directly on the ground because the warmth of the home will melt the upper layers of permafrost, and the building becomes unstable. The only way is to build on supports above ground so that air can circulate easily between it and the ground. Similarly, service pipes for a community such as water and sewage must be raised above ground within insulated containers in order to maintain the flow of whatever liquids are there and, at the same time, avoid heating the surface of the ground beneath the pipes.

There is an active layer above the permafrost that thaws and refreezes seasonally. Within this layer a variety of conditions are found, including frost heave in warmer weather. Frost mounds are common, that is to say, mounds that are composed of varying amounts of ice. One of the most spectacular of these is

Figure 3.6. A large pingo from the Arctic shores of the Beaufort Sea. The name comes from a native word for a conical ice-covered mound. This landform is found in permafrost areas where the ground is forced upward by pressures from below when very low temperatures cause cracks in the earth.

the pingo, an ice-cored hill that can be as much as 50 meters in height and hundreds of meters in diameter (Figure 3.6). More than a thousand of these have been identified along the coasts of the Beaufort Sea, particularly near river deltas. They are often covered with 10 or more meters of sand and silt. When a pingo melts, it first forms a summit crater, not unlike a volcano, and then as it collapses it becomes a shallow-rimmed depression.

PRUDHOE BAY RESOURCES

The cold northern shores of Alaska do not look like a place that once was a warm tropical environment, full of rich vegetation, but that is exactly what it once was, hundreds of millions of years ago. In those ancient times the organic matter was laid down that in time became the oil and gas reserves of today. The North Slope petroleum province, as it is called, includes the entire coast of northern Alaska, the continental shelves of the Chukchi Sea to the west, and the Beaufort Sea to the north. About half of this petroleum province, 200,000 square kilometers of it, lies offshore.

Oil and gas are being extracted from this area and moved by pipeline to the south coast of Alaska for onward transportation by ship. The permafrost terrain through which the oil has to be taken is a constant challenge to the engineering skills of those involved. Equally challenging is the problem of remediation when accidents occur. The Arctic environment is fragile, and the low temperatures of water ensure that pollutants remain in place for long periods of time. In 1989, the oil tanker the *Exxon Valdez* ran aground near Valdez, spilling 10 million gallons of oil. More than 5,000 kilometers of Alaska's coastline were contaminated, and all kinds of marine life were decimated.

The oil and gas reserves for the whole petroleum province, with a concentration in and around Prudhoe Bay, which amounts to 70 billion barrels of oil and 40 trillion cubic feet of gas, constitute one of the largest in the United States and represent about one quarter of the nation's production of oil. Not all of the reserves are immediately available because recoverable resources are always less than the total available. Some estimate can be made of the efficiency of this work by comparing it with other oil-producing states. The top 10, in descending order of production, are Texas, Alaska, Louisiana, California, Oklahoma, Wyoming, New Mexico, Kansas, North Dakota, and Utah. Among these, Alaska has the highest output per well, the smallest number of wells, and the second-highest total output.

Oil exploration in northern Alaska goes back to the time of World War I when there was a perceived shortage of oil for the navy. Extensive areas of land were reserved by the navy at that time. Renewed interest was stirred at the time of World War II and continued for a decade after that war. The critical phase of exploration, however, came in 1967 when the Prudhoe Bay oil field was discovered. From that time onward, stimulated by the Arab oil embargo of 1973, there was continuous exploration and production, and by 1977, the Alaska pipeline was completed. The Arctic environment and frontier setting make both exploration and extraction very difficult. The almost roadless terrain necessitates air transport for moving all equipment. Work is best done in winter. The summer's melting of the surface makes it difficult to move about and endangers the fragile environment.

The building of the Trans-Alaska Pipeline, which runs more than 1,000 kilometers from Prudhoe Bay to Valdez on the south coast, was one outstanding example of the challenge of living with permafrost. Special methods of construction had to be used. Part of the above-ground section of the pipeline had to be insulated and also supported by pilings driven deep into the permafrost to compensate for ground subsidence. Heating of the oil was essential to ensure easy and speedy flow to ocean ports in southern Alaska; and since all of this heating could not be done at one point, such as the source, supplementary sources of heat were needed along the length of the pipeline. Costs of these measures were enormous. Initial estimates for constructing the pipeline were $900 million. Final costs were $7 billion.

The petroleum industry is a vital part of Alaska's economy, contributing 70 percent of the state's gross product. At the present time exploration and production on the north slope are declining, and so the state's coffers are suffering. One reason for the decline is the need to move into more hostile and more

environmentally sensitive areas that require the use of additional expensive tech-
nologies. Coupled with these difficulties is the ongoing problem of rising temperatures, already a reality in Alaska. Decreased ice cover, changes in wind and
ice drift patterns, and the break off of previously shore-fast ice are just three of
the changes taking place. Maintenance of roads becomes a new cost as surface
melting increases. Additional sand and gravel need to be added regularly to
airport runways as well as to roads.

International demands and commitments by the U.S. government to reduce
greenhouse gas emissions and therefore reduce consumption of oil are a threat
to Alaska's economy. At the same time there are benefits from global warming.
The removal of overburden in open pit mining is greatly eased by climate warming as the surface soils are softened, and therefore costs are lowered. Costs of
fuel, extension of the mining season, and decreased maintenance costs of equipment are other advantages. Alaska has enormous quantities of mineral resources.
There are deposits of gold, silver, copper, and zinc. One Alaskan mine is the
world's largest producer of zinc. Coal deposits constitute 40 percent of U.S.
total coal resources.

GLOBAL WARMING

Industrialization over the last one and a half centuries increased production
of the so-called greenhouse gases, mainly carbon dioxide and chlorofluorocar-
bons. These gases occur naturally in the atmosphere, trapping heat that would
otherwise escape from the earth, but it was their greatly increased concentrations
that accelerated the present warming trend, giving rise to the now-familiar term
greenhouse effect. Scientists at the University of Fairbanks, Alaska, who have
been studying temperature conditions throughout the state, conclude that with a
global increase of about 2.5 degrees Celsius over the coming century, Alaska
will experience a 5-degree increase because of the nature of wind circulation
patterns around the world. No single problem is more troublesome for this state
than the anticipated additional increase of temperature.

Almost all global warming is due to burning fossil fuels such as oil, natural
gas, coal, and wood, but Alaska does not have to wait for the predicted climatic
changes. The evidence is already at hand: Since 1990 the amount of carbon
dioxide in the air is increasing rapidly, air temperatures are rising, and the extent
of sea ice in the Bering Sea has dropped by about 5 percent. Discontinuous
permafrost is thawing in some locations, sea ice is thinning, and seven of
Alaska's glaciers are showing an average reduction in thickness of 10 meters.
The greatest increases came after 1976. Forests, wildlife on land and in the sea,
and every kind of human endeavor are affected annually by these changes. When
permafrost thaws, the surface of the ground begins to resemble the karst topography as ground areas collapse. It resembles the carbon found in the eastern
parts of the coterminous United States. In Alaska it is known as thermokarst

terrain. All over the state evidence abounds of its past destructive influence: There are abandoned homes, offices, roads, and bridges. The useful life of highways and airports is about seven years before the ground has to be leveled and resurfaced all over again.

In the boreal forest that covers most of the state, wind is one of the biggest hazards, particularly in coastal areas where shallow root systems and poor soil make the trees vulnerable. The storms that bring the high winds are created by the collisions between cold Arctic air masses and warm Pacific ones, and any increases in temperatures in the Pacific intensify these storms. At times, large-scale tree blowdowns occur. There are some 50 million hectares of forest in Alaska, one quarter of which is commercial forest.

Biological agents respond to climate, and one insect, the black-headed budworm, one of Alaska's most damaging insects, periodically defoliates western hemlock and sitka spruce whenever the growing season temperature rises above average, a change that invariably allows this particular insect population to increase dramatically. In 1996, 1 million hectares of forests were damaged by this and other insects. When trees are defoliated, they catch fire readily, and fire takes a huge toll of the state's trees every year. About 90 percent of them are set by lightning strikes, and periodically—on the average, every 10 years—between 1 and 2 million hectares are destroyed in a single year. Much of this destruction is both normal and beneficial, natural cycles that enrich a variety of ecosystems. The focus of concern here is the large and unpredictable increase brought on by global warming.

Coastal Alaska has a large population of native people who live out their lives in a subsistence type of economy. They are heavily dependent on wildlife, fish, sea mammals, and land resources, but their habitat is a precarious one. Spring runoff and ice dams, especially on the Yukon-Kuskokwim river system and its tributaries, cause frequent flooding. High tidal ranges on the Bering Sea coast, as much as 5 meters, coupled with strong winds, cause havoc on the 78,000 square kilometers of treeless, tundra, deltaic landscape. Here, over one 30-year period in the second half of the twentieth century, 500 meters of exposed coastline disappeared. On the northern coast, retreat of sea ice from shore can destroy access to marine mammals, a major source of food for the residents of this area.

Mitigating Damage

There is a double challenge facing Alaska: minimizing destruction from volcanism and earthquakes as well as countering the effects of global warming. The U.S. Geological Service is the main agency on which Alaska depends for assessment and advance warning of earthquakes and volcanic eruptions. This agency has expanded its programs because of the increase in the number of deaths from volcanism in the last two decades of the twentieth century, more than in all the previous seven decades added together. Coordinated programs are now in place among the three main volcano observatories—Alaska, Cascades, and Hawaii. Each observatory shares with the other two its data on seismicity, hydrologic conditions, and activity during and between eruptions.

Efforts to mitigate the effects of global warming are concentrated at the University of Alaska, Fairbanks. There, at the Center for Global Change and Arctic System Research, under the sponsorship of the National Science Foundation and the U.S. Department of Interior, a team of specialists are at work assessing risk and launching educational programs to alert the community at large to danger. Much research has yet to be done, but some recommendations have already appeared based on findings to date. Regarding the huge areas of boreal forest, for example, I indicated that they are often beset with attacks from fire, wind, and insects. Plans are now in place to immediately harvest trees that have died because of these threats, before neighboring forests are affected, then market the wood locally and assess the best usage for the deforested land.

Wildlife management in all of its forms is critical in a state like Alaska where so many native people are dependent on this source of food. It is axiomatic that the numbers of caribou, just to take one example, will change as temperatures rise, so new monitoring and management programs are now in place. Transferring reindeer caught in places affected by ice-coated vegetation to ice-free territory is typical of the kinds of actions that have to be taken. On quite another front, preparation of sites prior to construction will save costs in the long run. In contrast to traditional practice, vegetation and organic soil are stripped at a construction site and the land left for five years before building to allow permafrost to thaw naturally.

Alaska is quite dependent on tourism and anxious to maintain a good image in the eyes of all its visitors. Under conditions of significant global warming, the state sees a longer summer season and a heightened interest in winter sports as other places in the world experience warmer weather. To sustain the interest of visitors, roads and structures used by them must be located away from floodplains, wetlands, and sites that have discontinuous permafrost. Infrastructure adversely affected by climate change must be repaired rapidly, say business leaders, or Alaska will acquire a reputation as a poorly maintained state with deteriorating facilities.

REFERENCES FOR FURTHER STUDY

Brown, J., et al. *An Arctic Ecosystem: The Coastal Plain of Northern Alaska.* Strouds-
    burg, PA: Dowden, Hutchinson, and Ross, 1980.

Brown, R.J.E. *Permafrost in Canada: Its Influence on Northern Development.* Toronto:
    University of Toronto Press, 1970.

Cas, R.A.F., and J. V. Wright. *Volcanic Successions Modern and Ancient.* London: Allen
    and Unwin, 1987.

Chapin, F. S., III, R. L. Jefferies, I. F. Reynolds, G. R. Shaver, and J. Svoboda, *Arctic*

Ecosystems in a Changing Climate: An Ecophysiological Perspective. San Diego: Academic Press, 1991.

Fisher, R. V., and H. U. Schmincke. *Pyroclastic Rocks.* Berlin: Springer-Verlag, 1984.

Francis, P. *Volcanoes.* Harmondsworth, UK: Penguin, 1976.

Peters, R. L., and T. E. Lovejoy, eds. *Global Warming and Biological Diversity.* New Haven, CT: Yale University Press, 1990.

Sheets, P. D., D. D. Gilbertson, and J. P. Garrec. *Volcanic Activity and Human Ecology.* London: Academic Press, 1979.

Simkin, T., and L. Siebert. *Volcanoes of the World.* Stroudsburg, PA: Dowden, Hutchinson, and Ross, 1981.

Tazieff, H., and J. C. Sabroux. *Forecasting Volcanic Events.* Amsterdam: Elsevier, 1983.

Williams, H., and A. McBirney. *Volcanology.* San Francisco: Freeman, Cooper, 1979.

4

HAWAII

The state of Hawaii is a chain of islands near the center of the Pacific Ocean. The eight main islands that form more than 90 percent of the state are, in descending order of size, Hawaii, Maui, Oahu, Kauai, Molokai, Lanai, Niihau, and Kahoolawe. There are also many smaller islands that in total amount to about 3 percent of the state's territory. The name of the island of Hawaii is a bit of a problem because references to it in books and news reports are often confused with the whole state. It is not only the biggest island; it is more than twice the size of all the other islands combined. Because of its location at the eastern end of the state, it is the youngest volcanic island and therefore the one where the most volcanic activity is taking place. To avoid the confusion associated with the name, I will use the term *Big Island*, a term frequently used by Hawaiians, when referring to this island.

Big Island has five volcanic mountains, approximately 30 kilometers apart, separated from one another by saddles formed from past lava flows. As the newest island, it has none of the coral reefs that are common on the far western, much older islands, and little of its coast has yet been affected by erosion. Maui, Molokai, Lanai, and Kahoolawe, which were once probably a single volcanic peak, are all now separated by distances averaging 12 kilometers and by shallow waters usually 500 meters deep. Oahu, third in size, is the location of the state capital, Pearl Harbor, the East-West Center, and the well-known tourist attractions of Diamond Head and Waikiki Beach. Kauai and Niihau stand apart from other islands. They are at considerable distances from their nearest neighbors, and the water in between is quite deep compared with most of the other islands.

GEOLOGICAL OVERVIEW

The Pacific plate's behavior was the key to Alaska's earthquake and volcanic activities. It is the same in Hawaii, but the way things happen is quite different. Hawaii is part of a large volcanic mountain range, most of which is beneath the sea, the islands forming the state being the youngest and therefore the highest and most visible parts. The whole mountain chain was formed by what some call a hot spot beneath Big Island in the earth's mantle. As the Pacific plate continued to move northward and then northwestward, first at a rate of 70 millimeters per year, then 86 millimeters, over the last 70 million years, magma from areas 60 to 170 kilometers deep in the asthenolith erupted continually at the location of the hot spot, creating, over time, the volcanic mountains that constitute the state.

The history of the Pacific plate's movements is told in the evolution of the Hawaiian-Emperor Chain, which has already been mentioned in Chapter 2, a dogleg series of volcanic mountains stretching 6,000 kilometers across the North Pacific from Hawaii to the Kurile Trench, a subduction zone close to the Kamchatka Peninsula. It also provides a concrete illustration of sea floor spreading. The dogleg part, the change of direction when the Pacific plate switched from moving northward and took a northwesterly course, was caused by collisions between the Indian subcontinent and the Eurasian land mass 40 million years ago. Volcanic mountains in the chain that are older than 70 million years were carried down into the Kurile subduction zone.

Hawaiian-Emperor Chain

This series of volcanic peaks consists of more than 100 individual volcanoes with a total volume of a million cubic kilometers. Most of them are below the sea. Their ages are progressively older and their heights lower as they move away from the hot spot (Figure 4.1). The history of any one of these volcanoes can now be better understood because a new one, Loihi, presently under active study, is developing under the sea 30 kilometers southeast of Big Island, its peak approximately 1 kilometer below sea level. Loihi, so named from a word meaning "long," stretches north and south, its southern end directly on the sea floor and its northern part resting on the submarine slopes of Mauna Loa, Kilauea (Figure 4.2), and Mauna Kea on Big Island. These are the only active surface volcanoes of the chain, so they provide additional opportunities for research.

The heights of some of these volcanoes can readily be overlooked because so much of them is below sea level. Take the two highest, for example, Mauna Kea and Mauna Loa, and note that they stand, like all the others, on an undersea platform that rises far above the general level of the sea floor. A single contour line representing 2 kilometers of depth can be drawn to encompass all of Kauai, Oahu, Maui, and Big Island. Even this measurement does not reach far enough.

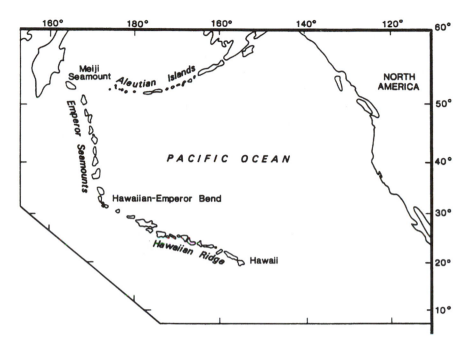

Figure 4.1. The Hawaiian-Emperor Chain is the name given to a series of volcanic islands stretching all the way from present-day Hawaii to the Kamchatka Peninsula. All these islands were created by a presumed "hot spot," permanently fixed beneath Hawaii, and the chain has formed over millions of years as the Pacific plate moves.

The average depth of the ocean floor in this part of the Pacific is approximately 5 kilometers. Thus the two highest volcanoes, Mauna Kea, 4,180 meters, and Mauna Loa, 4,144 meters, are each more than 9,000 meters above the ocean floor, higher than Mount Everest.

The identification of a hot spot anywhere on the surface of the globe provides one location that is fixed with respect to the earth's mantle. This is very valuable because all else is relative. All the plates are moving and have no fixed reference to the interior of the earth. It's the same with midocean ridges, oceans, and continents. These hot spots are found in a number of places all around the world, but their behavior over time is difficult to track if they occur on land. Yellowstone, for example, is a hot spot, but geologists are still trying to find out exactly where it erupted in past times. It is a much easier task when the spot is in the ocean, so the Hawaiian-Emperor Chain is an ideal model for studying the historical process.

Over the long 70 million years that we can trace this chain before the oldest of the volcanoes disappears beneath the Asian continent at Kamchatka, we can see a clear pattern developing for each individual volcano. First, as the Pacific plate moves westward and volcanic eruptions decrease, the rock that formed the

Figure 4.2. Kilauea caldera with Mauna Loa in the background. Kilauea's Iki crater is in the lower right foreground.

mountain gets cooler and cooler and therefore heavier. This increasingly heavy mass then presses down on the sea floor and pushes it downward. At the same time, erosive forces go to work on the top of the peak, first rain and wind, then ocean waves, until it is almost flat. The end result is a series of underwater volcanic mountains, each the same height above the sea floor, but with the flat top of each progressively deeper and deeper below sea level. The ocean floor had sunk, but each mountain ended up as a similar landform, often termed a *guyot*.

History of a Volcano

Because of their universal and multifaceted influence over conditions in the state of Hawaii, we need to be aware of the evolution of a typical Hawaiian volcano before examining its effects in daily life. Although the final appearance and size of a volcano will be unique, each one in the chain evolved and is evolving through the same sequence of stages. First comes the eruption of small quantities of lava deep in the ocean over the hot spot, gradually increasing in quantity until it reaches a peak about half a million years later, after which the amounts decline. Several million years may pass before eruptive activity finally ends and the volcano becomes extinct.

The volcano's life begins deep down below the ocean surface. Submarine eruptions build a steep-sided, small mountain with a shallow caldera. As the young volcano grows, small landslides cut into its steep slopes, scarring them. This first phase lasts about 200,000 years but produces only a small part of the final mass. An increase in the frequency and volume of eruptions marks the second phase, along with changes in the composition of the basaltic lava. This is Loihi's present stage of development, and before it is completed, Loihi will be close to the surface of the ocean, and explosive, ash-generating eruptions will become common as lava mixes with seawater (Figure 4.3).

The third phase is when the volcano has grown to more than a thousand meters above sea level and explosive eruptions begin to taper off. Lava flows are now low in volume and continue intermittently for several hundred thousand years. The type of lava emitted is shaped by the slope of the ground and the physical properties of the erupted basalt; most commonly, it is tholeiitic basalt. Lava that flows into the ocean shatters into sand and gravel-size fragments, and these blanket the submarine slopes. During all three phases the summit caldera repeatedly collapses, fills up, then collapses again. By the end of half a million years, more than 90 percent of the volcano's mass has been accumulated, and it looks like a warrior's shield—hence, the name shield volcano.

Weathering and erosion now take their toll on the high, steep-sided mountain. The seaward side, which is not supported as well as the landward, slips readily toward the ocean, creating large faults and causing major earthquakes. Occasionally there are catastrophic landslides. Recently vast fields of debris, some of it in large blocks, have been discovered all around the major Hawaiian islands. These submarine deposits suggest that major landslides must have occurred every 150,000 years on the average. Over time, deep canyons cut into the flanks, often along faults previously created by landslides.

At the same time, the volcano's enormous weight pushes the underlying lithosphere downward. Mauna Loa, for example, the world's biggest volcano, has a volume of more than 42,000 cubic kilometers. Thus the volcano begins to sink, and as this happens, fringing coral reefs grow at the shoreline, with sediments from the reefs accumulating in lagoons. In some Hawaiian islands, remnants of these ancient reefs can be seen. During times of global cooling when polar ice caps grew and sea levels dropped, the volcanic shorelines remained at the same level for long periods of time, allowing numerous large reefs to grow.

Beneath Mauna Loa, because of its enormous weight, the ocean crust is depressed approximately 500 meters, forming a hollow known as the Hawaiian Deep. In the adjacent waters the sea floor rises by about the same amount to form the Hawaiian Arch. Kilauea is 250,000 years ahead of Loihi in its evolution. It rose above the surface of the ocean 200,000 years ago and within the following 100,000 years had grown to a height of 700 meters above sea level. One of the interesting comparisons between Loihi and Kilauea is that both have three main areas of volcanic activity: the summit and two flanking rift zones. Though widely separated in time, it seems that their patterns of growth take similar paths.

Yet another stage, a rejuvenated one, can follow as much as 3 million years

Figure 4.3. Lava entering ocean at Kalapana, 1987.

later, during which strongly alkalic lavas are discharged, often through previously formed reefs. Ash cones are common features. Lava flows down the valleys that were carved out by streams during the erosional stage. Eruptions of this type, while their volcanic outputs are relatively small, have occurred on several of the Hawaiian Islands. The long duration of the rejuvenated stage on

Kauai and Niihau suggests that eruptions of this type could occur again on Maui, Molokai, and Oahu, or they could begin to erupt on Kahoolawe.

The last stage, as rejuvenation comes to an end, is more erosion and subsidence to the point where what is now an extinct volcano is worn down to sea level, then to an atoll with a ring of coral surrounding a lagoon. Coral growth may be destroyed through changes in water temperature or sea level, but if not, it forms a cap on the volcano. In time, as subsidence continues and it sinks below sea level, a guyot is the end result, a volcano flattened on top by wave and tidal erosive forces.

VOLCANIC ERUPTIONS

The Big Island consists of 5 volcanoes, as shown in Figure 4.4. Kilauea and Mauna Loa are the two that erupt frequently. Kilauea is located on the southeastern part of Mauna Loa. The eruptions of these two are in sharp contrast to the long, long process of volcanic evolution that I have just described. Overall, there are 15 volcanoes on the eight main islands ranging in volume from 6,000 to 42,000 cubic kilometers and from 390 to 4,170 meters in height. The 5 volcanoes on Big Island have the following volumes and heights:

	Volume in Cubic Kilometers	Height in Kilometers
Mauna Loa	42.5	4.17
Mauna Kea	24.8	4.20
Kilauea	19.4	1.25
Kohala	14.0	1.67
Hualalai	12.4	2.52

Kilauea

Kilauea, historically, has been the most active of all the Big Island's volcanoes. It is also the most intensively studied since the Hawaiian Volcano Observatory is on it, giving evidence by its presence of the mostly gentle nature of lava eruptions here when compared with some other places. No one would place an observatory close to the summit of Mount St. Helens. The Hawaiian Observatory was established in 1912 and became part of the U.S. Geological Service in 1948. It provides an excellent view of the Kilauea caldera (Figure 4.5). Its staff of more than 25 scientists conducts an ongoing program of geophysical, geochemical, and geological studies.

Within historical times Kilauea has erupted more than 30 times, and the most recent one, starting in the 1980s and continuing into the 1990s, was the longest lasting of all. The series of eruptions began on 3 January 1983, not unexpectedly because numerous minor earthquakes over the preceding days had confirmed

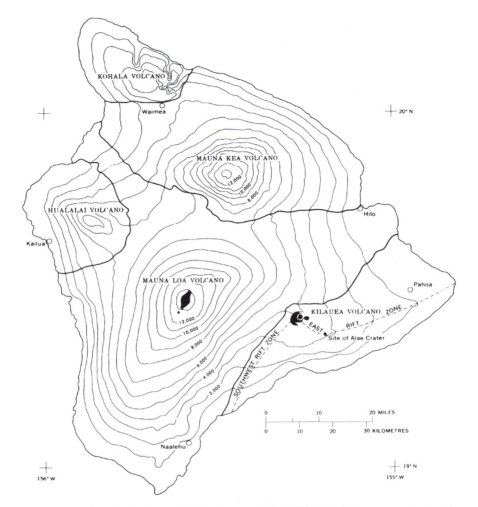

Figure 4.4. The Island of Hawaii, sometimes called the Big Island, is a group of volcanic peaks, and the contour lines show that the shapes of these peaks are somewhat conical. Kilauea has been the most active of the five throughout historical time. This island experiences most of the state of Hawaii's strongest earthquakes.

the imminence of a major event. By 2 January the quakes had risen to a strength two or three in terms of the Richter Scale. The lava first burst to the surface in a remote area on the flank of the volcano near an old crater on the rift zone, and the erupting fissures rapidly extended to a length of eight kilometers, producing in their wake a line of incandescent lava fountains appropriately named "a curtain of fire." Figure 4.6 illustrates this type of eruption.

All of Kilauea's eruptions seem to be fed from a single source, a magma chamber located 1,000 meters or more below the summit caldera. Rift zone

Figure 4.5. Aerial view of the Kilauea caldera and the Halemaumau crater. Mauna Kea is in the background.

eruptions receive their magma from dikes that branch out from the summit reservoir or from the conduits beneath it. Thus, when a rift zone eruption occurs, the summit of the volcano often collapses due to the draining away of the lava pools. Events of this kind were found to occur in Loihi when geologists from the Volcano Observatory examined it from a submersible.

Figure 4.6. The 1983 eruption of Kilauea.

The eruption of 1983 continued intermittently for a couple of weeks, then paused for about six weeks before beginning again. Speed of the lava was about six kilometers an hour as it flowed in the direction of the Royal Gardens subdivision, which fortunately was still undeveloped. Many more eruptive phases came in the months that followed, each one erupting from vents on the fracture system in the roadless and trailless fern forest on the middle east rift zone of Kilauea. Every fresh event produced spectacular lava fountains and flows toward the Royal Gardens. Lava flow speed was generally 1 meter per minute, with surges of 30 meters per minute. Everything in its path was totally destroyed, including 16 dwellings and over 300 lots in the subdivision.

Mauna Loa

Mauna Loa is much older than Kilauea. It has been active for at least half a million years and has a history of alternating rift and summit eruptions somewhat similar to those of Kilauea. In a typical pattern, Mauna Loa's eruptions are concentrated at the summit for several centuries, during which time the volcano's height increases. The summit then collapses to form a caldera, and a long series of rift zone eruptions is set in motion. Smaller eruptions occur at the summit in this period, but these are limited to some filling up of the caldera. A summit cycle then takes place again. The last two summit cycles continued for 1,500 and 2,000 years, respectively, and it is expected that another cycle will occur soon.

One year after the start of Kilauea's sequence, Mauna Loa began a long-

Figure 4.7. Fountains erupting from a fissure on Mauna Loa, 1984.

awaited major eruption. It began on 25 March 1984, climaxing 10 years of increased seismicity and accompanying growth in size of the volcano (Figure 4.7). There was considerable concern over the threat posed to the city of Hilo as lava first appeared at the 4,300-meter level. Later in that first day of activity, fissures opened up at lower levels, and by nightfall, most of the vents were concentrated at the 3,100-meter level. These vents produced 1.33 million cubic kilometers of lava per hour, flowing quickly downslope a distance of 14 kilometers in the first 24 hours. High-voltage power lines were snapped by the flow, and many square kilometers of native rain forest were destroyed.

The lava came within 8 kilometers of Hilo, but fortunately no further. The temperature of the lava remained at 1,140 degrees Celsius, cool for Hawaiian lavas, and while several of the channels collapsed, redirecting the lava flows, the danger to Hilo never rose beyond the 8-kilometer mark. In all, before the end of 1984, 325 million cubic kilometers of lava were erupted from Mauna Loa and 30 square kilometers of the volcano's surface buried. This was the first eruption of Mauna Loa for which detailed monitoring was possible, and much more was learned in the course of so doing than in all previous historic eruptions combined.

Mauna Kea

Mauna Kea is the highest volcano in the Hawaiian islands and is 900,000 years old. It represents that brief moment, geologically speaking, when a volcano reaches its apex after ages of eruptive activity and before millennia of erosion and decay tear it down. These forces of destruction are always at work: The underlying lithosphere sags as the height of the volcano increases, and this reduces its height; slumps and landslides over time remove much of the mountain and deposit it in the ocean; the same lithosphere that is heated and expands as it passes over the hot spot cools and contracts as it moves on and drags the volcano farther downward; there is also the cooling and sinking of the Pacific plate as it drifts away from the hot spot.

While a volcano is in the shield-building phase, it can grow faster than the destructive forces and therefore continue to grow. Mauna Kea seems to be at the end of that phase. Its last eruption was about 3,000 years ago, and while it will erupt again, it will be at a slower and slower rate of growth. Decay and erosion will increasingly be the dominant activity. The type of basalt erupting over recent millennia is more viscous than at an earlier age. Under these conditions the volcano is more and more steeply sloped, and the difference between the present outlines of Mauna Loa and Mauna Kea is therefore quite striking.

In relation to the other islands of the state, all volcanic, the pattern of the whole Hawaiian-Emperor Chain suggests that we will find the oldest and therefore the most eroded volcanoes farther west than the younger ones since they first erupted millions of years earlier. This is exactly what we find within the state of Hawaii on a smaller scale. Kauai is the remains of a single volcano 6 million years old. The island had plenty of erosion over its lifetime since the rainfall here is 12 meters per year, one of the wettest spots on earth. Deep gorges 1,000 meters deep mark the sides of the island.

OTHER HAZARDS

Earthquakes

Hawaii is one of the most hazard-prone locations anywhere on earth: hurricanes, droughts, rainstorms, flooding, shoreline retreat, storm surges, sea-level changes, erosion by wind, sea, and rain, plus the references already made to volcanic eruptions and earthquakes—all are experienced in this state. Earthquakes are common in the submarine parts of all volcanoes, but their strength is usually low. It is a different story when fault lines on the ocean floor become involved. More than 10 earthquakes of magnitude 5.3 or greater on the Richter Scale have hit Hawaii over the past 80 years, with all the biggest ones occurring near the multifaulted ocean floor beneath the Big Island.

November 1975 will long be remembered by Hawaiians. Early in the morning of that day everyone was awakened by a magnitude 5.7 quake that dislocated rocks on the south side of Kilauea, hurling them down a 300-meter slope. A party of boy scouts was camping close to this slope, and they quickly moved away and set up camp close to the shore. A little more than an hour later the whole island was shaken by a 7.2 earthquake that damaged or destroyed more than a hundred buildings. It also caused numerous landslides, damaged roads, and cut off electric power. All the scales on the island's seismographs showed maximum readings. Later estimates put the strength of the quake as a Richter 7.2.

Worst of all, a powerful tsunami was triggered by the earthquake. The scouts saw that the sea level was rising rapidly, but though they ran back toward the hillside, the incoming wave overtook them and knocked them down. The wave receded, but a second, much larger one soon followed. Many of the campers were thrown into a deep fissure, along with all kinds of debris. More waves followed. Piers, boats, houses, and vehicles all along the beach were destroyed. Later it was discovered that part of the coastal area just south of Kilauea had sunk three meters. Farther afield, on Catalina Island off California's coast, the tsunami was powerful enough to damage a dock.

Thirty minutes after the shock of the earthquake, signs of an imminent eruption began to appear. Soon a 150-meter line of fire fountains opened up on the floor of the Kalauea caldera, to be followed by two eruptions on the Halemaumau crater (see Figure 4.5). Later analysis concluded that a major block of the volcano had slid outward along a fault line 10 meters beneath the sea. An immense amount of the volcano had moved as much as 7 meters seaward. It was a huge landslide, and subsequent research showed that catastrophic collapses of this kind and even bigger had often happened here over geological time.

Erosion

Wind and, to a much greater extent, the ocean take their toll on the landscape. Much of the attractiveness of the state of Hawaii lies in the great sea cliffs and wide white sandy beaches that are the result of these forces. A combination of chemical decomposition of rock in the hot humid climate, minerals that are readily decomposed, and the powerful effect of saltwater waves accelerate erosion. Mass wasting, or landslides, is a frequent occurrence as erosional forces eat into the bases of slopes. In 1999, disaster overtook a party of tourists on the north side of Oahu as tons of rock slid down the mountainside.

Soils in any volcanic environment are excellent, especially where the climate is warm so that ash and basaltic lavas can be weathered quickly. This is why people in many parts of the world who live in volcanic areas choose to till the soil on the slopes of the volcanoes. In some countries two or three crops a year are possible in such soils. In Hawaii, before the development of tourism and the presence of military installations, there was a rich agricultural industry. It still is a significant part of the economy, but soil erosion is taking a heavy toll of topsoil, and attempts are being made to limit the loss to five tons per acre per year. At that level, new soil can replace what is lost. This goal is still elusive. In Kauai, much more than five tons per acre is presently being lost annually.

Hurricanes are rare, but when they do come, they bring all the destructive power that we associate with tropical storms. Hurricane Iniki struck Hawaii on 11 September 1992, causing extensive damage to property and coastal areas. It was the most powerful hurricane to strike the Hawaiian Islands in a century. Kauai experienced winds of 225 kilometers per hour and was the most extensively damaged island. Storm surges there extended inland as much as 300 meters, rising to heights of 9 meters in places. The estimated cost to Iniki was $2.4 trillion.

Tropical storms have winds ranging from 63 to 118 kilometers per hour, and storms are classified as hurricanes if they have winds greater than 118 kilometers per hour. Between 1960 and 2000, 68 tropical storms and 42 hurricanes hit Hawaii, their frequency being greatest when the ocean temperature is highest. For the most part, the islands are small targets in a big ocean, so landfalls are rare. Over the 100 years prior to the arrival of Iniki, only five other hurricanes are listed as major. Iwa, which struck on 23 November 1982, was the closest to Iniki in amount of damage, about one tenth of Iniki's. All the others when added together only amount to a small fraction of the destruction caused by Iwa.

Tsunamis

It has sometimes been said that Pacific Islanders, unlike most of humanity, live on the ocean and occasionally visit land. Whatever truth may be in this statement, it is certain that one visitor that always lives in the ocean is never welcome in Hawaii, a tsunami, a Japanese word meaning "Great wave in harbor." By its position in mid-Pacific, Hawaii experiences tsunamis from all the countries that border the Pacific. In the almost 200 years since records were first kept, the Hawaiian Islands have been hit with 100 tsunamis, an average of one every 2 years. Occasionally a Hawaiian earthquake is big enough to cause one of these, but for the most part, the threat is related to earthquakes elsewhere.

The source of most Hawaiian tsunamis is South America, and the second biggest is the Kamchatka Peninsula. Both of these areas experience subduction as the Pacific plate moves under the adjacent continents, and we know that the most violent earthquakes, and therefore the biggest tsunamis, will occur there. Alaska, which is the third biggest source of Hawaiian tsunamis, is also the place where the power of a subduction earthquake was dramatically demonstrated in 1964. Japan is number four. Thereafter, there are several countries, all at the same level of numbers, where Hawaiian-bound tsunamis originate. They include Mexico, California, Indonesia, and the Solomon Islands.

A tsunami is defined as a train of progressive, long waves generated in the ocean by an impulsive disturbance. That disturbance may be caused by a submarine volcanic explosion, through a landslide in which a block of land plunges into the water, or by an earthquake on the ocean floor. Great earthquakes often have their origins under the sea, especially near the shores of the Pacific Ocean where huge tectonic plates meet. There is no certain way of knowing if a tsunami will follow an earthquake, but it is commonly accepted that only earthquakes of magnitude 7 or more are likely to cause destructive tsunamis.

When the ocean floor is raised suddenly or dropped during a major earthquake, a huge mass of water is suddenly set in motion, and this water either rushes away or rushes in, causing a lot of sloshing backward and forward for hours. This is the beginning of a tsunami. Its velocity increases with its wave length, which in turn is determined by the depth of the ocean, and in the deep ocean, the wave length can be very long, creating a speed of 800 kilometers per hour with wave intervals of one hour. Once the epicenter of an earthquake is known, and the ocean depth also known, it is possible to predict the arrival time of a tsunami by calculating its speed. That speed is the square root of the ocean depth multiplied by a factor of approximately four; thus at 2,000 meters depth, speed will be 500 kilometers per hour, and at 200 meters, 150. The height of the wave may be only 30 centimeters above the general level of the ocean, and so it is scarcely detectable until it reaches shallow water.

As the wave approaches land, its bottom part is slowed down by friction as it encounters shallower water, while the top experiences much less friction and catches up with the bottom. For some sea floor topography and orientation, the wave can rise up to a height of 9 meters or more and rush onto shore as a wall of water, causing enormous destruction. For a different topography the same wave may have a much less pronounced effect, leading merely to a surge and then a withdrawal of water with only minimal damage to shore installations. A tsunami can travel 20,000 kilometers across the Pacific at great speed and still cause enormous damage at its destination.

Two memorable tsunamis live on in the memories of Hawaiians, the Alaskan of 1946 and the Chilean of 1960. The 1946 one happened on 1 April, All Fools Day, a factor that cost valuable time for some, thinking that warnings were a joke. The quake hit Alaska at two o'clock in the morning, and the tsunami that followed sped outward across the Pacific. Hawaii was directly in its path, and though more than 3,600 kilometers distant, it was only five hours away at the speed of tsunamis.

The biggest changes in sea floor depth come from shallow subduction quakes, the kind that are common on the west coasts of North and South America, the kind that hit Alaska in 1964. The enormous change in the sea floor during that earthquake produced a series of tsunamis. Some of them battered several Alaskan communities, particularly Kodiak Harbor, where a wave nine meters higher than high tide smashed onto the shore. At the same time, waves began to move oceanward at right angles to the axis of the uplift.

When the tsunami hit Big Island at 7:00 in the morning, most islanders were up, fortunately, and able to act quickly in response to warnings. Some young people on their way to school were fascinated with the way the huge waves receded and left new areas of ocean floor bare before returning with fresh fury to wreak havoc on land. During that in between time, many of the boys and girls ran to the seawall to see the sight. Suddenly, too quickly for most of them, the next powerful wave arrived. Some ran to higher ground; others clung to trees and shrubs as water crashed over them. A few were lost and never seen again.

The town of Hilo on the northeast coast of the island of Hawaii suffered an enormous amount of damage. Its location facing the direct path of the tsunami ensured that. Its bayfront business district in Hilo Bay was almost totally destroyed. Almost every house on the bay side of Main Street was smashed against buildings on the opposite side. The Hilo train station disappeared. Rails were torn off their beds and in places wrapped around trees. A steel bridge across the railway near the bay was torn off its supports and washed a quarter of a kilometer upstream. About 100 residents of Hilo lost their lives.

A combined total of 60 lives were lost on four other islands. Altogether 500 homes or businesses were totally destroyed and a further 1,000 severely damaged, adding up to property losses of $26 million. We need to keep in mind that these figures are based on 1946 values. It would be a very much larger amount today. Roads, railroads, piers, bridges, and ships all were affected. Waves reached a height of 17 meters in one valley and elsewhere swept four fifths of a kilometer inland. Between crests, the sea floor was bare for 150 meters beyond low tide.

The second memorable tsunami came from Chile in 1960, the result of a series of powerful quakes that began on the morning of 21 May. The first earthquake triggered a weak tsunami that did no more than generate a small wave on one of the Hawaiian islands. Chile is much farther away from Hawaii than is Alaska, over 10,000 kilometers, compared with 3,500 kilometers from Alaska, so 15, rather than 5, hours is the time required for a tsunami to travel from Chile. It was the third quake that generated the fateful tsunami. It arrived at Hawaii in the early morning hours of 22 May.

This time Hawaii was well prepared. Two years after the 1946 tragedy an elaborate warning system was installed. It included observation posts throughout the Pacific Ocean so that initial indications from tidal gauges could be noted and the information relayed to Hawaii or any other island nation affected (Figure 4.8). The very moment that a seismic disturbance severe enough to cause a tsunami is now observed anywhere in the Pacific Basin, it is recorded at the Honolulu Observatory. Epicenter location and times of arrival at affected shores are estimated and warnings passed on. Other nations supplement this information. The Japanese Meteorological Agency, for instance, has five tsunami forecast centers, and these also give advance warnings.

Unfortunately a change in the way the Hawaiian warning signal was broadcast by radio left islanders puzzled. A decision had been made to move from a three-stage signal to a single one, but many people had forgotten about the change and so waited for a second and third siren. As in 1946, the town of Hilo on Big Island suffered more severely than anywhere. Other islands experienced only moderate damage. At Hilo Bay the highest wave towered more than 10 meters above normal sea level and raced inland at 50 kilometers an hour. Boul-

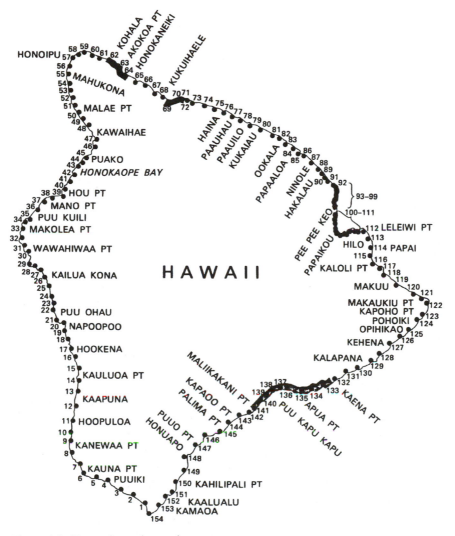

Figure 4.8. Tsunami warning stations map.

ders as heavy as 20 tons were picked up from the bayfront seawall and carried 180 meters across a park without leaving a mark on the grass. Five-centimeter pipes supporting parking meters were bent over parallel to the ground. Entire city blocks were swept clean, buildings being wrenched from their foundations and deposited as piles of debris 100 meters away.

Over 500 buildings representing $50 million of value were destroyed. At Hilo alone, 61 people were crushed or drowned by the waves, and an additional 43 required hospitalization. The sewage system was not functional because its pumps were out of action and its pipes broken. For some time everyone had to boil water before use, and electricity was unavailable for a long time. Public shelters were employed to house the 215 families who lost their homes.

NATURAL RESOURCES

Energy

With so much volcanic action in its environment, you would expect Hawaii to make substantial use of geothermal energy, as is done in several countries including the coterminous United States. One feature of Hawaii's volcanoes, the intrusion of magma into underground spaces in the existing rock and consequent creation of a network of dikes, provides an excellent setting for geothermal or hydrothermal systems. These dikes give protective insulation for the magma, allowing it to cool much more slowly than when eruptions occur on the surface. In these circumstances heat may be absorbed by groundwater and discharged as steam or hot water.

Because geothermal systems depend on recent magma intrusions, the younger volcanoes are the ones that can provide a continuous flow of heat. On Kilauea, a geothermal system with temperatures exceeding 300 degrees Celsius has been identified at depths of two kilometers, extending along the entire length of its east rift zone. Other islands have geothermal systems, too, but with lesser capacities. These are not the only possible energy sources. Both high temperatures and high winds suggest abundant resources for developing solar and wind power. Additionally the sharp drop into deep water from many of the volcanic peaks makes it possible to use the differences between water temperatures at surface and at depth to generate power. Several experiments have been conducted with a view to making that source economically viable.

Sadly the reality is very different from what we might expect. It seems to be cheaper to import energy than to try to produce it locally. Throughout the last quarter of the twentieth century, oil has contributed to well over 80 percent of primary energy consumption, and about one third of that oil comes from Prudhoe Bay, Alaska. The following statistics illustrate the ongoing dependence on oil, a dependence that is greater than in any one of the other 49 states. It is evident that some moves are being made to develop geothermal energy and to use waste materials to generate electricity, but the biggest change from dependence on oil is the introduction of substantial quantities of coal in the mid-1990s, hardly an addition that will improve the quality of the environment.

Primary Energy Consumption in Hawaii (in percentages of the total)

1975	1985	1995
Petroleum 99	Petroleum 90	Petroleum 87
Others (mainly solar)	Biomass 9	Coal 5

1975	1985	1995
	Others (mainly solar) 1	Biomass 4
		Solid waste 2
		Geothermal 1
		Solar 1

Water

The volcanic nature of the islands ensures one major benefit—abundant supplies of freshwater. There is no need for desalination plants. The mountainous nature of the island chain provides more than 7,000 millimeters of mean annual rainfall. Were it not for these peaks, Hawaii would have only the ocean-level mean, about one tenth of what is presently available. While adequate supplies are available on most islands, there are some desert areas. Traditional agriculture depended on perennial streams originating in the mountains.

Favorable geologic conditions is the reason for the preservation of large quantities of water. Otherwise, rainfall would be lost through surface runoff into the sea. Soils and rocks on the surface allow both easy infiltration and subsurface movement of water because the thin-bedded basaltic lavas that are found almost everywhere on the islands have many fractures and open spaces through which water can easily flow. At a deeper level, thick, massive lavas that are much less porous act as storage places for water. Dikes, volcanic intrusions that cut vertically across lava flows, act as barriers to water flows so that water is often impounded by them, forming aquifers.

The largest bodies of freshwater are known as basal groundwater—that is to say, bodies of freshwater that float on saltwater within an aquifer. Basal water bodies lie no more than a few meters above sea level. The greater buoyancy of the freshwater allows them to extend below sea level by as much as several hundred meters, depending on whether or not it is confined by a caprock, a thick sequence of coastal plain sediments. These sediments are derived from muds and other fine-grained material and therefore have very low permeability. Below the basal water is a transition zone of mixed fresh- and saltwater, followed by the main mass of saltwater (Figure 4.9).

Basal water bodies are recharged in two ways, from direct infiltration of rainwater and from underground flows from higher elevations. The latter is the bigger source, the former being a variable and uncertain supply. Overall the amount of water stored in an aquifer depends on three things: the permeability of the rock through which water flows, the amount of recharge water received, and the presence or absence of some confinement such as caprock. In addition to the dominant groundwater resources found in basal water, there is high-level freshwater that does not rest on seawater. Both of these groundwaters are the preferred domestic water supplies because of their purity.

The single biggest problem in water supply is the uneven distribution within islands and from island to island. Oahu withdraws 375 million gallons per day

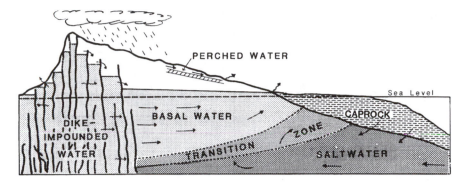

Figure 4.9. Water supplies for the Hawaiian Islands come from impounded quantities of fresh rainwater. Vertical, impervious, volcanic dikes store large quantities of this basal water, exerting pressure that prevents saltwater encroachment. Some rainwater is caught above horizontal rock layers and perched above sea level.

from groundwater; Maui, 200; and Hawaii, 50. Other islands take less than 50 each from their underground resources. Population in the resort centers tends to concentrate on the dry coastal areas where rainfall and recharge are at a minimum. Overall there is sufficient water to meet all needs, but in particular locations, there are looming crises. Oahu's demands, for example, are running close to maximum availability, and the steadily increasing population will force either a change in consumption patterns or the necessity of finding new supplies.

Mitigating Damage

In the aftermath of the 1960 tsunami, it was recognized that Hilo would always be at the mercy of a big tsunami, so much of the bay front was converted to recreational and parking spaces. The ocean-side buffer zone was widened and a landfill plateau created to provide an area of higher ground. On other islands preventive measures were also taken. Houses built on stilts seemed to fare better that those built directly on the ground, so this discovery shaped design in new construction. Reinforced concrete structures were least affected by waves. Hotels now have open ground floors with high ceilings. This tsunami-resistant design assumes that ground floor and basement could be flooded. The concrete structural columns on the ground floor are designed to withstand the impact of the waves, whereas the nonstructural walls between the columns are regarded as expendable.

Increased densities of people and services on shorelines pose increased risks of damage from tsunamis. It is obvious that there are economic benefits from locating transportation and commercial facilities close to major water routes, just as there are aesthetic pleasures in living by the sea, but these things must be evaluated against the dangers to life and property. With a steady increase in coastal population densities year after year, the likelihood of major tragedies occurring continues to rise. Throughout Hawaii, whenever there is an imminent risk of a big tsunami, people are urged—and if necessary, forced—to move to higher ground and keep clear of tidal areas. From past experience it is well known where the limits of reach are, for even the biggest historical tsunamis, so safety is at these higher inland spots.

I have already described the warning system used in Hawaii to alert the population when a tsunami is likely to strike. Similar systems operate in Alaska, California, and Japan. These are vital components of mitigation procedures. However, there are many other steps that can be taken in advance to minimize damage: constructing breakwaters and seawalls, together with planting of protective forests to safeguard places that are close to the shore; locating such facilities as land and air transportation terminals and combustible and contaminating storage tanks out of the hazard areas; designing buildings to withstand or divert wave forces; and using the power of zoning to locate buildings where people usually gather in large numbers, as in schools and hospitals, outside the hazard areas.

REFERENCES FOR FURTHER STUDY

Austin, P. C. *An Environmental Overview of Geothermal Development in Hawaii.* Honolulu: Department of Planning and Economic Development, 1979.

Ayre, Robert S. *Earthquake and Tsunami Hazards in the United States.* Boulder, CO: Institute of Behavioral Science, 1975.

Claque, D. A., and D. McKenzie. *Tertiary Pacific Plate Motion Deduced from the Hawaiian-Emperor Chain.* Boulder, CO: Geological Society of America Bulletin, Vol. 84, 1973.

Cox, D. C., and J. Morgan. *Local Tsunamis and Possible Local Tsunamis in Hawaii.* Honolulu: Hawaiian Institute of Geophysics, 1977.

Department of Geography. *Atlas of Hawaii.* Honolulu: University of Hawaii Press, 1973.

Hazlett, R. W. *Geological Field Guide, Kilauea Volcano.* Hawaii: Hawaiian Volcano Observatory, 1987.

Heliker, C. *Volcanic and Seismic Hazards on the Island of Hawaii.* Honolulu: Bishop Museum Press, 1993.

MacDonald, G. A., A. Abbott, and F. Peterson. *Volcanoes in the Sea.* Honolulu: University of Hawaii Press, 1970.

Stearns, H. T. *Geology of the State of Hawaii.* 2nd ed. Palo Alto, CA: Pacific Books, 1985.

Stearns, H. T., and G. A. MacDonald. *Geology and Ground-water Resources of the Island of Oahu.* Honolulu: Hawaiian Division of Hydrography, Bulletin 1, 1935.

U.S. Geological Survey. *Natural Hazards on the Island of Hawaii.* Washington, D.C.: Government Printing Office, 1976.

5

CORDILLERAN OROGEN

The Cordilleran Orogen is a mountainous region stretching from the Pacific to the Great Plains and from Canada to Mexico, covering the states of Arizona, California, Nevada, Oregon, Utah, Washington, plus parts of some other states. Its geological history is tumultuous. It's the part of the nation that always comes to mind whenever earthquakes and volcanic eruptions are in the news. It is the place where all the different kinds of seismic activity are found: subduction and strike-slip earthquakes are here; volcanic eruptions and continental migration are also here. All the types of earth movements that we have seen in Alaska and Hawaii can be found in the Cordilleran Orogen or on its borders.

The geology of the whole region can best be understood as three zones: (1) Cascade Range, the series of volcanic peaks in the Northwest; (2) San Andreas Fault, the strike-slip fault running through central California and having many branch faults; and (3) the Basin and Range Province, a place of distinctive volcanic activity that coincides with the physiographic region of the same name. I am going to take these zones of seismic activity in the order in which I have listed them. Before beginning, I want to recount a fascinating piece of geological detective work. It will be a useful introduction to the first zone, the Cascade Range.

Throughout historical time, there is no record of an earthquake of magnitude 8 or 9 at the Juan de Fuca site, yet all along the Pacific Coast of North and South America, there have been earthquakes of this magnitude in all the other regions. Is the Pacific Northwest unique? Do the gigantic Pacific and North American plates behave differently here than they do everywhere else? It was the remaining portion of an older Pacific spreading ridge, namely, the Juan de Fuca plate, that caused the massive 1964 earthquake in Alaska, but that same plate is subducting beneath the Pacific Northwest. Why are there no big earthquakes there? These questions troubled geologists for a long time.

By the mid-1980s, refinements in our knowledge of plate tectonics, particularly awareness of the much greater power in plates that are subducting when the distances from their spreading ridges are relatively short, persuaded scientists to search for evidence of past quakes in the Northwest. Perhaps, they speculated, there were powerful earthquakes in the past of which we know nothing because this part of the country has been settled for such a short time. One development spurred these scientists on: A nuclear power plant was about to be installed in the Seattle area, and regulatory authorities wanted to know if there were any seismic concerns that ought to be taken into consideration. Field study near Seattle began in earnest.

Digging down beside a stream close to the coast a geologist found layers of sand and mud below the surface extending downward for a couple of meters before coming to an abrupt stop at a junction with a layer of peat. As he dug down farther he found another layer of sand and mud below the peat. It was obvious that this part of the coast had once been below high water, then above it, and then below it again before coming to its present state with topsoil in the uppermost 20 centimeters. Furthermore, the sharp demarcation lines between peat and sand suggested that the changes from below to above water had been sudden, just the sort of thing known to be typical of subduction earthquakes.

If this was indeed the result of a former subduction earthquake, there ought to be similar layers of mud and peat at the same depths all over the same place. Before long, evidence of such was found in abundance. In one place a number of additional sets of layers were uncovered, suggesting that there might have been a succession of subduction quakes, separated by long periods of quiescence. As you moved farther and farther back from the coast, looking carefully at each location where layers of mud and peat were found, a very fine layer of sand on the peat seemed to become thinner and thinner, the further you were from the coast. This is exactly what you would expect if a powerful tsunami had swept across the land following an earthquake.

Samples of plant material and bits of wood from the top peat layer were collected and their age calculated using the carbon-14 technique. They were found to be approximately 300 years old. At the same time all sorts of additional data kept coming in, all confirming the original speculation that this region has always experienced massive subduction earthquakes, each separated from the next by several hundred years, the last one being before there was any European settlement, about 300 years ago. It was at this stage in their investigations that the geologists made contact with a Japanese seismologist who happened to be visiting North America and took an interest in the Pacific Northwest's research.

This seismologist was well acquainted with subduction earthquakes and the tsunamis that so often accompanied them. He concluded that if one had occurred here about 300 years ago, it ought to be possible to prove that it had happened. Japan, being an older civilization, has records of earthquakes and tsunamis going

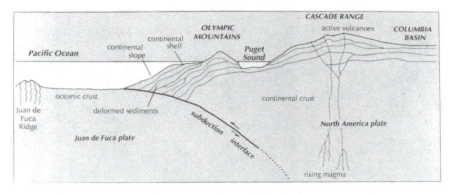

Figure 5.1. A cross section of Washington State showing the Juan de Fuca plate being subducted. As the ocean crust pushes downward, parts of the North American plate are pushed upward, thus giving rise to volcanic activity as cracks in the rock allow magma to reach the surface.

back many hundreds of years. When he returned to his country, he found that a powerful tsunami, of the kind that would be triggered by a magnitude 9 earthquake, had struck Honshu, Japan's main island, exactly 300 years ago and that it came from this part of North America. Knowing the speed of the tsunami, he was able to say exactly when the earthquake took place. It was 26 January 1700.

CASCADE RANGE

The long history of subduction activity is the key to the history of the Cascade Range. These mountains are located approximately 200 kilometers from the coast. If we were to imagine a line descending directly downward into the earth from these mountains, we would encounter the subducting ocean crust at a depth of about 100 kilometers. Ocean crust from the Juan de Fuca ridges descends slowly, so it has traveled some distance inland before it gets down to 100 kilometers (Figure 5.1). At that level, as was explained in Chapter 1, the heat in the crust and its associated water and other volatile material are high enough to create the pockets of magma that rise close to the surface. If a conduit permits any of this magma to break out onto the surface, we have a volcanic eruption.

The Cascade Range is a series of stratovolcanoes, that is to say, volcanic mountains that are conical in shape because of the nature of the lava that built them over the past thousands of years (Figure 5.2). Other volcanoes such as those in Hawaii have a different shape because the lava that built them has a different chemical composition. The Cascades run from Mount Garibaldi north of the Canadian border, all the way to Mount Lassen in northern California, and each peak has its own unique history; some have a very violent past, whereas others either took shape quietly or we do not know enough about their past. We

Figure 5.2. Mount Hood, one of a chain of mountains, the Cascade Range, that runs from the Canadian border all the way to northern California. The shape of the mountain shows that it is the result of volcanic activity and also that this activity is recent, geologically speaking, since erosion has not had time to change the mountain's shape.

will look briefly at some of these volcanic peaks, then study in greater detail the 1980 massive eruption of Mount St. Helens.

Mounts Lassen, Mazama, and Rainier

Beginning at the southern end of the Cascade Range, at Mount Lassen in northern California, we have the place where, apart from Mount St. Helens, the most recent major eruption occurred. That was in 1915, with continuing action at different times over the following two years. The type of rock erupted was mostly basalt, but in the explosive action accompanying the main event, hot rocks were thrown onto the snow-covered flanks of Mount Lassen, triggering major debris flows. In Lassen Volcanic National Park, which includes Mount Lassen and three smaller volcanic centers, the present-day geothermal system has a magma pool near enough to the surface to provide hot water. This system consists of a reservoir with a temperature of 235 degrees Celsius underlain by a reservoir of hot water.

Crater Lake, farther to the north in Oregon, is the remnant of a catastrophic eruption of what was once Mount Mazama, 3,600 meters high (Figure 5.3). In that gigantic upheaval of 7,000 years ago, about 50 cubic kilometers of matter was flung into the air in a pyroclastic flow that spread ash over large parts of eight states. It was an explosion that far exceeded that of Mount St. Helens,

Figure 5.3. Crater Lake, Oregon, site of an ancient, catastrophic eruption that occurred almost 7,000 years ago.

probably 10 times greater in power and volume. As often happens in an event of this magnitude, the summit of the mountain sank back into the now-empty 8-kilometer-wide inner magma chamber and went down below ground level. The empty caldera so formed then filled with water to form the place we now know as Crater Lake.

Mount Baker has a summit rising to more than 3,200 meters. After Mount St. Helens it is the most active volcano in Washington State. Steam and gas emissions from it were common during the 1970s. Glacier Peak is a volcanic mountain very similar to Mount Baker. Its height is more than 3,100 meters, and, like Baker, its cone is less than 1 million years old. Mount Hood is the fourth highest peak. It has been active ever since the middle years of the Miocene period. Unlike most of the lower 48 states, the northern Cascades has a concentration of alpine glaciers, a remnant of the Holocene Ice Age, and this is responsible for the rugged topography of the area.

Mount Rainier is the highest mountain of the Cascades, towering more than 4,300 meters above sea level. It has the greatest concentration of glaciers of any mountain in the lower 48. Like other peaks of the Cascades, its growth occurred within the last million years, but within that period, all sorts of volcanic eruptions, landslides, and mud flows devastated the surrounding area. As recently as 1989 a gigantic rock avalanche of 2 million cubic meters of rock crashed down the north flank of the mountain. It is these debris flows from the glaciers that have given a great deal of concern to surrounding settlements.

Glacial outburst floods originate when water stored at the base of glaciers is suddenly released, and floods of this kind have been launched from four of Mount Rainier's glaciers. The most prolific of the four is South Tahoma Glacier, which had 15 of these outbursts between 1986 and 1992. These floods occur during periods of unusually hot weather in summer or early fall. Rainy weather can also be a trigger. I used the term *debris flows* in relation to these because the release of water triggers small landslides and picks up sediment on the way down. This flow of water, mud, and rocks at ground level then travels at about 20 kilometers an hour, tearing up vegetation and damaging roads and facilities in Mount Rainier National Park.

Mount Rainier is receiving a lot of attention these days. Part of the concern relates to its rock structure, the presence of weak layers of rock high on the mountain, as well as its huge cap of snow and ice. These areas could collapse in even a small earthquake and disrupt life and industry in nearby Seattle. During the 1990s, it was selected by the U.S. Geological Service for intense study as one of three places that might cause major damage over the next decade or two. To emphasize the urgency of this study, it was pointed out that a fault line, known as the Seattle Fault, runs from a point near Mount Rainier to Seattle. Given the frequency of smaller earthquakes in and around Seattle, there is every justification for concentrated research efforts while, at the same time, keeping a close watch on the mountain's behavior.

Mount St. Helens

Mount St. Helens is the youngest of the Cascades' volcanic peaks, and the explosion of 1980, referred to briefly in Chapter 1, is just the most recent of the many intermittent eruptions that have taken place over the past 40,000 years

Figure 5.4. Mount St. Helens in 1983, three years after the cataclysmic event of 18 May 1980. Ash and gas emissions can be seen rising from the 200-meter-high dome that developed in the crater over the previous three years. The present size of the crater is three kilometers in diameter. The earthquake explosion of 1980, measuring 5.1 on the Richter Scale, initiated one of the largest landslides and eruptions ever recorded in the coterminous United States.

(Figure 5.4). Pumice and ash from these past events now cover large areas of the Pacific Northwest. From the 1950s onward, the mountain was intensively monitored, perhaps to a greater degree than any other. Days before the fateful event of Sunday, 18 May 1980, there were many signs of impending danger, but no one was quite prepared for what finally happened, the largest eruption in the history of the coterminous United States.

It all seemed to take place in seconds. Seattle's air traffic control tower tracked the mass of ash and rocks hurtling out of the mountain and concluded it was traveling at close to 500 kilometers an hour. The earthquake that triggered the explosion measured 5.1 on the Richter Scale, but the energy released might be more accurately defined as the equivalent of thousands of Hiroshima-size bombs. An avalanche of mud, rock, and ice roared down the mountainside, while the ash cloud rose as high as 18,000 meters. What had moments before been a beautiful 3,000-meter-high mountain was reduced to a 2,500-meter decapitated peak.

More than three cubic kilometers of material was moved in those first few minutes. Ash, high in the atmosphere, drifted eastward right across the country, covering the ground everywhere it went with a layer of ash and blocking out sunlight in several communities near at hand. Six hundred square kilometers of forest was flattened. Mud flows rushed westward down river valleys toward the Columbia River, blocking the navigation channel for ships with logs and mud for a distance of 15 kilometers. It was estimated that 57 people lost their lives on that first day.

A Boeing 737 jet flying from Reno, Nevada, to Vancouver, Canada, at 10,000 kilometers altitude was 60 kilometers south of Mount St. Helens when it exploded. The pilot saw the explosion and swung away from his course, a path that would have taken him directly over the eruption, escaping in so doing a dirty gray cloud that was rising quickly to meet him. His 737 rocked in the air from the shock of the explosion as if it were a ship at sea. Fortunately his flight had been delayed a half hour at Reno, or all 122 passengers plus crew would have been added to the list of deaths for 18 May.

David Johnston, the expert geologist from the U.S. Geological Service, who was monitoring the mountain on the morning of the explosion, talked to news reporters early on that day. He described Mount St. Helens as a keg of dynamite with a lit fuse whose length you do not know. He was well aware of the risks of being so close to the summit, but he stayed on there right up to the moment of the eruption. He told the reporters that it was extremely dangerous to stand where they were at that time. "If the mountain explodes," he told them before they left, "we would all die." Soon afterward they heard his final words as he yelled into this radio, "Vancouver, Vancouver, this is it!"

A family watched their dream $100,000 home smashed and washed down the chocolate-brown Toutle River. A young couple were on a camping and fishing trip on the same river when they heard the explosion. They tried to grab their camping gear but quickly saw there was no time to escape in their car. They were thrown into the water and carried along in a mass of mud, logs, and rocks, grimly clinging to one log. They were lucky. The log was shunted sideways out of the main stream. Sometime later, a helicopter picked them up. A television cameraman was a kilometer and a half from the base of the mountain, filming the event. He saw the mass of mud and debris heading for him, so he dropped everything, got into his car, and drove madly to keep ahead of destruction. He was able to stay ahead.

Farther east, travelers were stranded in numerous small communities, altogether 10,000 of them in three states. One couple driving west from Spokane saw the black oncoming cloud. Soon they could only inch along the highway at less than eight kilometers an hour. They abandoned their car and joined the other stranded ones. Everywhere around trains, buses, airplanes, and cars came to a stop. Walking was the only thing that worked. Digging out from under the ash was yet another hazard. It proved to be as hard as getting around it. For some time people wore masks of whatever material they could find for fear of toxic fumes.

As is common in stratovolcanoes, the magma that had risen and caused the explosion of 18 May left the inner chamber empty for a time until new magma moved up from below. The interior dome then began to grow until pressure rose to a level that caused another eruption. Several of these subsequent eruptions came in May, June, July, August, and October of 1980. By 1983 the dome had grown to 200 meters, and the crater in which it sat was three kilometers in diameter, waiting for the moment of the next event and meanwhile continuing to grow.

How can volcanic eruptions and earthquakes be predicted? The answer remains elusive, but experience at Mount St. Helens shows us some of the things that can be done. As I mentioned earlier in this chapter, this mountain had received more monitoring since the 1950s than almost any other, and the small number of people who were killed is largely due to this as well as the actions that were taken in the months before 18 May 1980. The first earthquake in Mount St. Helens struck on 20 March 1980, and immediately seismologists met with local authorities to warn of the danger ahead and make some preliminary plans.

One week later steam and ash exploded from the summit of the volcano, and this was followed by several minor eruptions over the following weeks. As these eruptions became more frequent, public authorities closed off the area around the mountain, causing heated opposition from the general public. Later they lowered the water level in the Swift Dam reservoir to minimize damage from mud flows. Still closer to the eruption the state governor issued a state of emergency in order to use the National Guard to staff the roadways. So angry were some over the closure that they found ways of circumventing the law by using little-known roads and footpaths to gain access. Many of these people were too close to the volcano when it exploded and died. Some, like Harry Truman, a veteran resident of the mountain, refused to leave his home on the north side and died.

WASHINGTON STATE

The western portion of the state is an accumulation of millions of years of crust, some of it from erosion processes and sedimentation, some accreted to the state from island sources in the Pacific. Farther east lie the oldest sedimentary and metamorphic rocks, and these contain some of the most valuable sources of mineral deposits. Magnesium, dolomite, and magnetite have all been economically mined here. Zinc, gold, silver, cobalt, and uranium deposits are also mined, the last named being the cause of some concern as erosion into younger sediments generates radon gas, a hazard to humans, one that we will examine in more detail in a later chapter.

Floods

Volcanic eruptions and earthquakes are not the only environmental issues facing Washington State. This is an area of high rainfall, and frequently when

Figure 5.5. The Yakima River and gravel pits near Selah Gap, Washington, after the flood of February 1996. The center front of the picture is where a dike had been. It was breached, and the gravel pits were flooded, with the result that long-term damage was done to the salmon habitat.

there is heavy precipitation, the rain affects mining activities at lower elevations, causing flooding or serious erosion. Floodplains are regularly mined to provide sand and gravel for construction projects, but sadly these activities do not take adequate account of potential changes in rainfall and environment.

I have already referred to this problem in my introduction in connection with 100-year floods and the tendency for industry to ignore events that might be a century away. In Washington, *floodplain* is defined in government regulations as the 100-year floodplain, that is, the area that has a 1 percent likelihood of flooding in any given year. That 1 percent risk can become a 100 percent risk at any time, not just 100 years after the last big one. One reason for this lies in the unpredictability factor as watersheds become more and more developed and roofs and roads create large areas of impermeable surfaces, greatly increasing the danger of flooding.

From late December 1996 to early in January 1997, Washington experienced a series of winter storms that deposited abnormally high precipitation. The culprit was a cold continental air mass that sat over northwest Washington while a series of warm fronts moved in from the Pacific Ocean. On the Yakima River, as shown in Figure 5.5, dikes were breached, and water flowed into a series of gravel pits. One serious outcome of this event can be seen if we consider part of the life cycle of salmon, one of the major economic assets of this part of the country.

It is easy to forget that rivers are not isolated ribbons of water but rather are connected systems of tributaries, surface water flows, and shallow groundwater sources. During spring and fall, juvenile salmon migrate out of the main channels of rivers into these smaller streams to escape from the high water. While they are within these tributary channels, they take advantage of a rich supply of aquatic insects and at the same time avoid the predators that occupy the deeper water courses. When an avulsion like the one in the Yakima River breaks into these habitats, the resultant turbidity and erosion destroy both the salmon and their habitat.

Landslides

Landslides are another common hazard in the environment of the Northwest and elsewhere on the west coast, so much so that the president of the United States declared parts of coastal Washington to be a disaster area following the heavy rains of December 1996 and January 1997 (Figure 5.6). Buildings on coastal bluffs were particularly vulnerable.

The reason for this vulnerability dates back to the last ice age when poorly consolidated sands were deposited in fractures and gullies beneath the surface rock of impermeable till, which is a strong and resistant cap extending down 10 meters or more. The problem with these sands is that they rest on a slippery surface of perched groundwater so that with each passing winter, they slide some distance down the slope, creating the stepped profile that is a characteristic feature of this coastal region (Figure 5.7).

Columbia Plateau

This upland area, the basin of the Columbia River, occupies a very large area of eastern Washington and extends into southwestern Idaho and northern Oregon. It is characterized by incised rivers, extensive plateaus, and steep ridges. There are some areas where minerals such as clay and tin can be economically mined, and in the southern parts, limited quantities of natural gas were extracted in the years before World War II. During the long stretches of time between periods of volcanism, forests sprang up and spread, only to be enveloped later by new flows of basalt. Quantities of petrified wood can now be recovered from these times, and the Washington State legislature has designated petrified wood as the state gem.

The long history of volcanic activity (Figure 5.8) now forms a cover of Miocene basalts extending over more than a third of the state of Washington in depths that frequently approach 5,000 meters.

Figure 5.6. Looking west over the Magnolia Bridge, a major artery into downtown Seattle, Washington. In the months of December 1996 and January 1997 a series of storms caused flooding and landslides in many parts of the Northwest, including the area shown in this photograph. The bridge was closed, and five homes on the bluff in the background were declared uninhabitable. Notice the displaced bridge support.

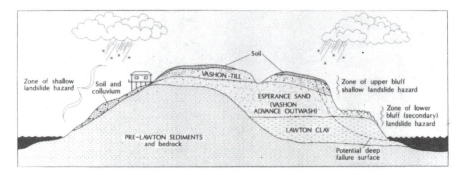

Figure 5.7. Typical stratigraphy in the Seattle area where landslides occur. The Vashton Till is resistant to erosion and largely impermeable. However, cracks allow water to penetrate to the Esperance sands below and these, when sufficiently wet, slide along the surface of the Lawton Clay onto the bluff.

Last Ice Age

The Columbia plateau has been extensively studied because these great depths of a consistent layer of volcanic rock seemed to be ideal places for depositing nuclear wastes. During the Pleistocene period, gravel, sand, silt, and clay were deposited on the surface, in places accumulating to depths of 300 meters. Glacial outwash also produced huge volumes of loess, that is to say, windblown sand deposits. There are other landforms, too, and their number and complexity are evidences of the permanent marks on the landscape left by the last ice age.

The Columbia Basin, for example, was the scene of the greatest series of floods ever documented. As the Pleistocene ice sheet advanced south into Idaho, it dammed a major river at the border of Montana to form a lake equal in volume to that of present-day Lake Michigan. Later, as different sections of ice began to melt, the ice dam gave way, and water cascaded down the Spokane Valley and onto the Columbia Basin, greatly modifying the surface of the land. Flow rates were greater than the total flow of all the rivers of the world today. Channels were cut through the loess and other surfaces, leaving a jumbled topography of many different types of landforms known collectively as Channeled Scablands.

These things happened because lobes of ice pushed southward from western Canada, forcing their way along river valleys and lowlands. These lobes were part of a continental ice sheet of great thickness that advanced under the pressure of a growing mass of ice behind it. Local glaciers in the mountains contained much less ice, and they moved downslope by gravity. The mechanics of these two very different ice flows led to the variety of landforms. The land was sculptured by glacial meltwaters and climatic changes. Rivers were diverted from their normal courses, huge inland lakes were formed, and extended periods of frozen ground left the landscape looking like places we see today in Alaska.

Figure 5.8. Portland, Oregon. Two hazards are always present in this city: (1) the volcanic activity that might erupt along the chain of active mountains to the east and (2) the flooding that can occur when snowmelt from the upper reaches of the Columbia reaches Portland.

SAN ANDREAS FAULT

The San Andreas Fault is the single most important element in the interaction between the Pacific and the North American plates, and some of its effects are felt far inland across the western part of the country. It is a rare situation to find two of the biggest tectonic plates, the Pacific and the North American, meeting on land at the San Andreas Fault. This fault, a strike-slip one, moves as much as five centimeters a year and has been doing this for more than 50 million years, with a total displacement of hundreds of kilometers. Massive earthquakes, like the 1906 one that we will examine in a moment, are associated with it or with its numerous associated faults.

Both the Pacific and North American plates are moving relative to the deep mantle, so the San Andreas Fault boundary is also moving, changing its shape in the process as the adjacent plates deform. In southern California, the sector of the fault from north of Los Angeles to east of San Bernardino has been rotating slowly counterclockwise. We tend to think of the margins between these plates as narrow lines because this is the way they are depicted on maps, and we also tend to imagine the rest of the Cordillera as being static. The reality is quite different. In places the plate margin may be hundreds of kilometers wide, and the whole of the Cordillera may be in motion at different rates in different places.

San Francisco Earthquake, 1906

Early in the morning of 18 April 1906, about 5:12 A.M. while most people were still in their beds, the big earthquake hit San Francisco with a strength that was subsequently estimated at 8.3 on the Richter Scale. If we compare its magnitude with that of Loma Prieta, which we will look at in a moment, it had 30 times the power of Loma Prieta. It lasted only 48 seconds, but that seemed like a year to those who were rudely awakened and had to rush out into the streets with whatever clothing they could lay their hands on. Aftershocks soon followed, and the devastation they could see in every direction convinced most people to stay away from their crumbling homes.

Earthquakes had struck California before. In 1812 an area south of Los Angeles was hit, and more than 30 people were killed by it. In 1857 another one caused considerable damage over an area northeast of Los Angeles. Some decades later a geologist from the University of California, who had observed a fault line south of San Francisco, decided to trace its extension north and south. He and his students found, to their surprise, that this was no ordinary fault. It was the San Andreas Fault, and it ran almost the full length of the state, close to the coast, for the most part, but veering inland in the south. This happened in the 1890s, long before there was any understanding of plate tectonics, so little was done with the new information.

It took the 1906 event, the first major assault on a big city by an earthquake, to set in motion a serious quest for the cause of the earthquake. About 12 square kilometers, almost 500 city blocks, were devastated by the quake. Masonry buildings collapsed, but wood-frame homes and skyscrapers withstood the shock. One exception was the landfill areas in Marina District near the water. Wood-frame homes there just disintegrated. Electrical power lines, water mains, and all the other normal services were cut off. When the first fire broke out, nothing could be done about it because there was no water.

The worst horror came later in the morning with fires all over the city, 60 in all. A firestorm erupted to add to the terror. The fires raged for three days, with a total destructive power 20 times that of the earthquake. The city was incinerated. All attempts to create firebreaks failed. There were probably 3,000 or more killed, or 1 percent of the population, but the number of dead was never released in case businesses decided not to return. When the fires finally subsided, people searched for their homes or what might be left of them. It was a difficult task. All the familiar landmarks had vanished.

The quake was felt over an area close to half a million square kilometers. Cities close in suffered varying amounts of damage. Stanford University was one of the worst affected. Several buildings were completely destroyed there.

There is a reason for the extensive harm done to cities nearby: San Francisco has the highest density of faults of any urban area in the United States. The break in the San Andreas Fault that caused the 1906 disaster lay 10 kilometers below ground, and the amount of lateral displacement was as much as six meters in places. Costs in 1906 dollars were close to $500 million. That would be about $7 billion today, a figure close to the cost of the 1989 Loma Prieta earthquake.

First attempts to understand the behavior of the San Andreas Fault began within a year of the 1906 disaster, first by a geologist who had lived through the quake. He recognized that the cause of the earthquake was slippage on part of the San Andreas Fault, later identified as a segment that stretched for 400 kilometers from Monterey Bay northward. Pipelines and roads that crossed the fault line had been broken and displaced by an average of four meters, with the western side always moving northward with respect to the eastern side. It was evidently a strike-slip fault, but until the era of plate tectonics, everyone regarded it as an anomaly, a one-of-a-kind event unique to California.

A few years later, in a final report on the 1906 earthquake, an engineer who examined land surveys of the area around San Francisco from the nineteenth century and compared them with those taken after the 1906 earthquake made an interesting observation. Landmarks that lay some distance to the west, such as a lighthouse 40 kilometers offshore, kept moving northward over time with respect to landmarks to the east. He concluded that this relative movement between land on two sides of the big fault must lead to tension, rather like stretching a piece of elastic material. At some point the tension becomes so great that one side of the fault snaps as it moves to relieve the strain, and an earthquake occurs. That's what happened in 1906. Now we wait for the next one. It will surely come, somewhere near the San Andreas Fault. Meanwhile, geologists continue their difficult task of trying to predict when that will be.

Loma Prieta Earthquake, 1989

In 1989 a 7.1 magnitude earthquake, the Loma Prieta, ruptured a 40-kilometer segment of the San Andreas Fault zone in the Santa Cruz Mountains about 85 kilometers south of San Francisco. It was the largest to strike this part of California since 1906 and was felt over an area of a million square kilometers, from Los Angeles to the border of Oregon. Destruction included 67 deaths, 3,800 injuries, and more than 18,000 homes damaged. Losses as a result of the earthquake amounted to $6 billion. The segment of the fault that was ruptured had been recognized for some time as having the greatest probability of any of the faults associated with the San Andreas for producing a major earthquake.

Thousands of landslides generated by the quake were found all over an area half the size of the one hit in 1906. Loma Prieta thus provided the first opportunity to study the effects of a major earthquake on landslides (Figures 5.9–5.11). Previous landslide-producing earthquakes, apart from the 1906 one, were either too small or too poorly documented for this purpose. Techniques for

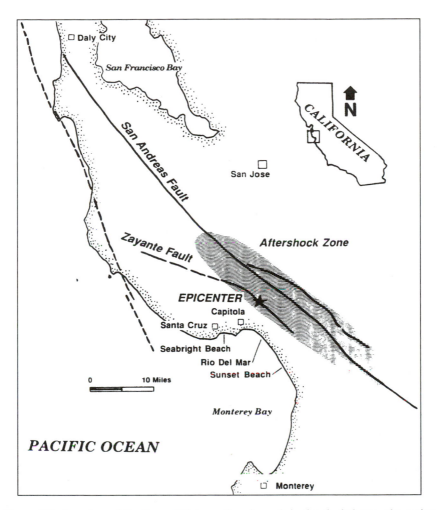

Figure 5.9. Location of the Loma Prieta earthquake and, in the shaded area, the region affected by aftershocks. The quake struck along a 40-kilometer segment of the San Andreas Fault and was the largest to hit central California since the one in 1906. The earthquake was felt all the way from Los Angeles to the Washington-Oregon border.

identifying slopes susceptible to failure that had been developed over the previous 10 years were proved correct in the studies that followed the Loma Prieta earthquake. At the same time there was recognition of new types of landslide hazards not fully appreciated in the past, particularly those associated with fault traces.

The losses associated with this earthquake were fortunately small considering the high population density and the numerous vulnerable seismic conditions in the Loma Prieta region. The damage that did occur was typical of landslides in

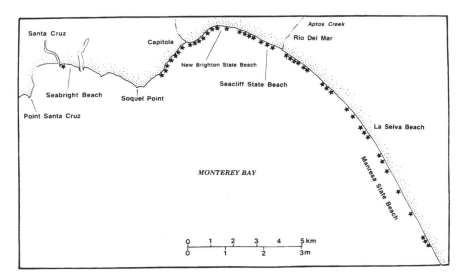

Figure 5.10. All along the coast near the epicenter there were a series of coastal-bluff failures as a result of the Loma Prieta earthquake. These cliffs are about 30 meters high, the long-term result of coastal erosion and uplifting of land. The cliffs are mainly sandstone, and the uppermost layers are poorly consolidated sand and gravel.

other parts of the state or, for that matter, anywhere in the nation: long-term disruption of roads and highways and delays in recovery due to the continuing instability of earthquake-damaged slopes.

CALIFORNIA

The geologic history of central California can be traced back to the Jurassic period when a volcanic island arc was probably attached to the western part of North America in the foothills of the Sierra Nevada. At the same time, the Sierras were uplifted. Subduction of ocean crust then began along a new plate boundary west of the accreted volcanic arc. Sea floor spreading ferried all kinds of rocks, sediments, and terranes to the area of the subduction trench, forming what we now know as the eastern part of the Coast Ranges.

In the northern coastal areas of California the Coast Ranges are seen to be almost one continuous mountain chain. Farther south they form two or three parallel ranges with intervening alluviated valleys. The highest peaks in these southern ranges are 1,700 meters high, and the valleys can reach widths of 20 kilometers. Between these two mountain masses, the Coast Ranges and the Sierra Nevada, lies the Great Valley of California, a fertile alluvial plain 650 kilometers long and up to 90 kilometers wide. The west slope of the Sierra Nevada is incised by streams that flow into the Great Valley to augment the

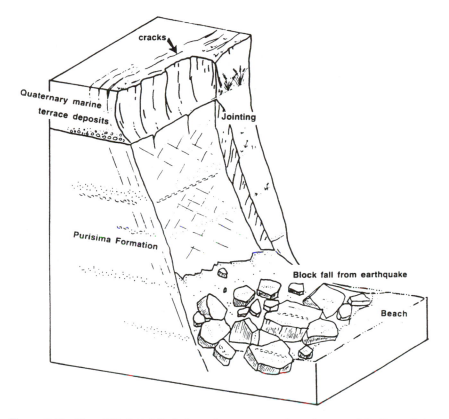

cracks

Quaternary marine
terrace deposits

Jointing

Purisima Formation

Block fall from earthquake

Beach

Figure 5.11. The cliffs that failed along the coast near the epicenter of the Loma Prieta earthquake often gave way in blocks of poorly consolidated material that capped the sandstone. The process was aided by undercutting of the upper layers and by the presence of tension cracks in these same deposits.

Sacramento River in the north and the San Joaquin River in the south, both of which flow northward into the San Francisco Bay and thence to the Pacific.

The climate of California is Mediterranean. The dry season is May to October, and the wet season, fed mainly by polar front cyclonic storms, is November to April. Orographic lifting intensifies precipitation on the western slopes of the Coast Ranges, whereas adiabatically warmed descending air leaves the eastern slopes relatively dry. In the San Francisco Bay area, for example, 150 centimeters of rain is common on the coastal slopes of the Coast Ranges, whereas only 25 centimeters falls on the eastern side at the same latitude. Environmental issues are numerous and complex: storm-induced damage when unusually heavy rainstorms occur; coastal erosion; subsidence as a result of the withdrawal of groundwater; and widespread landslides whenever there are earthquakes or heavy rains.

Subsidence

We will take the last-mentioned first, subsidence due to withdrawal of groundwater. No other state draws such a large portion of its freshwater supplies from underground. While land subsidence can be caused by all kinds of things—mining, earth movements due to faults, compaction of sediments and unconsolidated formations, and earthquakes—the reality is that the largest volume of land subsidence in the world due to human activities took place in the Great Valley of California. More than one half of the San Joaquin Valley, an area of 13,500 square kilometers, subsided by more than one third of a meter on average and by as much as nine meters in places. Some action was taken in the 1970s to correct the situation by cutting back on withdrawals.

Whenever groundwater is withdrawn beyond a certain amount, seawater migrates into the freshwater aquifer. There is always a transition zone between fresh- and saltwater areas. The latter is generally held back by the hydraulic head in the freshwater aquifer. If this head is lowered, saltwater advances, and this can destroy agricultural areas. In the 1930s, for example, at a time of low rainfall, heavy pumping on the east side of San Francisco Bay and in the Santa Clara Valley created a wide area of seawater intrusion, and at the same time, significant land subsidence occurred.

Coastal Erosion

All of the coast from Los Angeles to San Francisco experiences moderate erosion all the time. There are several reasons for this. For one thing, conflicts between people and nature always exist along the coasts because people have a persistent desire to live and work on the coast. California is one state that led the way in largest increase of coastal population over the past two decades. I mentioned this problem in my introduction as one of the human-caused environmental hazards. As more and more people move to areas that are unstable, any destructive event in these places puts a lot of people at risk.

The mainland beaches of California border an active tectonic region of crustal blocks and high elevations. Ancient shoreline terraces hundreds of meters above present sea level provide proof of rapid and extensive crustal uplift along the coast. The weather is temperate year-round, and the prospect of a cliff-top home overlooking the ocean can be so enticing that the danger of cliff collapse is ignored. More than once, as we saw in the cliff failure at Seattle, winter storms take a heavy toll on these elevated shores.

The reason for the severe pounding lies in the nature of the beaches. Submarine canyons are very close to shore, and so there is a very narrow and steep zone beneath the cliffs that receives the brunt of the wave action. The only source of sediment for building up the beaches is the material eroded from headlands and carried along the coast in longshore drift. However, the submarine canyons usually interrupt this longshore movement of sediment and carry it out

Figure 5.12. The landslide undercut the foundation of this California home, removed the sidewalk, and threatened the road in the foreground. Of all the geological hazards facing the nation, landslides are probably the most common. California, because of earthquakes, often experiences more than most states.

to sea into deep water. Landslides are common, as I have said, whenever winter storms hit the coast, but landslides are even more widespread farther inland. Indeed, the two major types of landslides, the storm-induced and the earthquake-induced, are probably the most destructive of all of California's environmental problems.

Storm-Induced Landslides

The first five months of 1995 were examples of the destructive effects of heavy rainfall. Overall, precipitation was five times the normal amounts, with some areas, such as Santa Barbara, experiencing quantities that exceeded all past records. Ten people died as a result of either flooding or landslides, and total damage to agriculture and buildings amounted to $1.5 billion. The winter storms that brought all this rainfall began early in January, causing widespread flooding and thousands of landslides, debris flows, and road failures throughout the state (Figure 5.12).

One landslide in Ventura County occurred on a known active landslide that was activated by the storms. Elsewhere, as in Humboldt County, the land gave way in locations that were known to be susceptible to failure. Public awareness after the events of situations like these where precautionary action could have been taken created demands for a more rigorous study of terrain conditions. The staff of California's Division of Mines and Geology responded in three ways: (1) Emergency response teams were set up for northern and southern areas; (2) communication links were established between these teams and all other agencies of the state concerned with environmental emergencies; and (3) programs of field reconnaissance were planned for all areas known to be high risk.

One danger associated with landslides relates to streams or rivers that are dammed by the slide to form a lake. In Mendocino County in northern California, such a lake was formed during the storms of March 1995. For a time all was well. The flow of water in the river concerned, the Navarro, was reduced to a fraction of its usual volume, while the lake above the dam grew steadily in width and depth. No industrial or domestic operations were affected by the reduced flow of water. The big problem arose when the pressure of water breached the dam, and suddenly an abnormal amount of water rushed downstream in a short period of time. Frequently the damage caused by these burst dams is worse than the destruction inflicted by the original landslide.

Earthquake-Induced Landslides

Large parts of California, especially the coastal stretches from San Francisco southward, have a history of fatal and destructive landslides that are triggered by heavy rainfall, coastal and stream erosion, human activity, and earthquakes (Figure 5.13). The last-mentioned is our focus here. The great San Francisco earthquake of 1906 generated more than 10,000 landslides over an area of 32,000 square kilometers, killing 11 people and causing widespread damage. In the San Francisco–Monterey region over the past 150 years, there have been at least 20 other earthquakes, and each has caused a number of landslides. The conditions that make this area particularly vulnerable are the steep and rugged topography, weak rock and soils, and active seismicity.

BASIN AND RANGE PROVINCE

East of California as far as Colorado (Figure 5.14), Utah, Nevada, and parts of several neighboring states is the Basin and Range Province. It is inland, far from the junction of the Pacific and North American plates, yet here are some of the most powerful examples of seismic activity to be found anywhere on the continent. I mentioned earlier that we should not think of the Cordillera as a static mass of mountains and plains but rather as a place where earth movements of many kinds create conditions for earthquakes. Fault scarps no older than a few thousand years are present in many parts of the province and prove the existence of past earthquakes.

Figure 5.13. San Francisco. Landslides and earthquakes are never far away in time from this part of California. The San Andreas Fault passes close to the city on its western side. The photograph shows the Golden Gate Bridge with the city of San Francisco in the foreground and the Bay of the same name in the center.

Long Valley

In May 1980, one week after the eruption of Mount St. Helens, a strong earthquake swarm occurred at Long Valley near the eastern boundary of California, about 50 kilometers northwest of Bishop. This area is at the western edge of the Basin and Range Province and has become the latest focus of attention by the U.S. Geological Service. A huge eruption occurred here 700,000 years ago, and because no repetition of that has taken place, there has been a tendency to discount the possibility of danger from volcanic activity. A considerable amount of lava dome construction did occur about 500 years ago, but that, too, seemed too remote an event to cause concern today.

Now with the latest series of earthquakes, together with earlier developments in and around the Long Valley caldera, this region has become the third place, along with Hawaii and the Pacific Northwest, to receive renewed attention from the U.S. Geological Service. There are good reasons for this fresh concern. A hundred years ago there was an eruption on one of the volcanoes on the 3,500-meter-high Mammoth Mountain, just west of Long Valley. In October 1978, a magnitude 5.7 quake jolted the town of Bishop. Then came the swarm of earthquakes in 1980. Still later, further swarms came in the early 1990s.

Figure 5.14. Flooding is a common experience in all parts of the United States, but flash flooding in the steep canyons of Colorado is unusually destructive. Here, in the aftermath of the 1976 Big Thompson Canyon flood, a battered truck is buried in flood-transported soil and rock debris.

As concerns rose over the number of earthquakes, it was discovered that the inner area of the Long Valley caldera was rising. Uplift of a caldera floor is one of the sure signs of a long-dormant volcano coming back to life because it means that magma in the chamber below is being forced upward by new magma from deeper down. The question in the minds of geologists now is this: Are we seeing a repeat of the kind of thing that happened at Mount St. Helens where a series of earthquakes led up to the final explosion? This uncertainty is the explanation for all the attention that is being devoted to the Long Valley caldera.

Grand Canyon

Farther east in the Basin and Range Province, about 500 kilometers southeast of Long Valley, the Grand Canyon winds its way westward. Few geological sites can match it as a model of the long history of erosion, uplift, and deposition that characterized this central portion of the Cordillera. These processes, in all kinds of sequences, are etched in the sides of the canyon, making it a popular site for tourists. So popular is the Grand Canyon National Park that there is no parking space any longer if you happen to arrive at a weekend or holiday in the summer. From the park's point of view, there is a more important concern, the steady destruction of the natural habitat by all the cars, motor homes, sightseeing flights over the canyon, and the 5 million annual footsloggers.

In 1999 decisions were taken to banish most motor vehicles and flights and replace them with a light rail and bus system and a network of walking and bike trails. Behind the decision is a desire to make the park a model for other parks throughout the nation. It is to take effect within three years, and as we would expect, it is receiving applause from environmentalists. While this rehabilitation of the park is going ahead, preserving its flora and fauna and reducing noise levels, another ambitious project aimed at the restoration of living conditions for plants, animals, and fish within the canyon has just come to fruition. It aims to restore and maintain the historic habitats of life in and at the margins of the river while still using the dams on the river to generate electrical power.

Before Hoover and Glen Canyon dams were built, the Colorado flooded every spring, carrying large quantities of sediment that accumulated on the sides of the river and at junctions with tributary streams where the speed of the water was slowed down. At low water these deposits of sediment became beaches. A large number and variety of life forms were sustained by this cycle of low and high water. With the dams came control, so water flowed evenly year-round. Flooding, which had been a problem in some low-lying areas, disappeared, and new supplies of electricity were being generated by a nonpolluting source. It seemed at the time a perfect solution to two problems.

The perfect solution was not so perfect when the river's habitat is given adequate consideration. Sediment that was transported at flood times was now trapped in the upper reaches of Lake Powell. Water that was warm in summer was now cold all year long because it came from deep in Lake Powell. The results: The canyon lost its sand deposits, and five of the endemic species of fish disappeared because the water was too cold. New species of fish that appeared were detrimental to those native to the river in its original state. To correct the damage done, periodic flooding was instituted because it was evident that the absence of floods did serious damage to this flood-adapted ecosystem.

The results of the switch to the river's natural state, or as close as we can get to it, have been hailed as completely successful. Sandbars are restored, backwaters and riperian vegetation reestablished, and native fishes have a greater competitive advantage. Secretary of the Interior Bruce Babbitt stressed the sea change in the operations of large dams. They can be operated for environmental purposes as well as water capture and power generation.

UTAH

The rich variety of landforms and challenging environments in the eastern Cordillera are well represented in the state of Utah. Overall, three of the big physiographic provinces of North America take up most of the land surface of the state—Great Basin of the Basin and Range Province, Wasatch Range and

Uinta Mountains of the Rocky Mountain Province, and much of the Colorado Plateau Province. The last mentioned, in the southeast corner of Utah, contains the Canyonlands, sheer-walled canyons and cliffs, plateaus, mesas, buttes, and badlands carved out of sandstone and limestone rock by the Colorado River and its tributaries. West of the Canyonlands is a high plateau marked by a series of escarpments.

In the north of the Colorado Plateau, toward the western flank of the Rockies, are the Uinta Basin and still farther north the Uinta Mountains with some peaks exceeding 3,000 meters and with the mountains as a whole extending 50 kilometers wide and stretching east-west for almost 250 kilometers. These mountains were extensively glaciated during the Pleistocene, so glacial features are evident everywhere—horns, aretes, cirques, and glacial troughs. Ground and terminal moraines filled the valleys after the last ice age, damming rivers and creating hundreds of lakes. The Uinta Basin is within the Colorado Plateau Province. It has some deeply cut ravines and many canyons similar to those of the Colorado. It also holds ancient marine limestone and sandstone deposits, some as thick as 8,000 meters, containing vast deposits of oil shale and other hydrocarbons.

The Wasatch Range in the northern part of the state trends north and south, extending from Salt Creek Canyon into Idaho. The higher parts of the range were glaciated in Pleistocene times, so the same glacial marks that occur in the Uintas are also found here. In addition to sedimentary layers, the Wasatch Range has intrusive igneous rocks. These deposits account for Bingham Canyon's copper wealth as well as for significant lead, silver, and zinc resources. The western side of the Range is very steep, the result of displacement along the extensive and still-active Wasatch Fault (Figure 5.15). This is the same fault that runs through a densely populated part of Salt Lake City and stands as the greatest earthquake hazard anywhere in the state.

Thistle Landslide

The year 1983 was Utah's wettest of the century, and it followed an extremely wet 1982. Large areas of the country were equally affected, and the culprit was another El Niño cycle. For Utah, a state with normally low levels of precipitation, the two very wet years caused extensive damage from landslides, flooding, and debris flows. Many old landslides started moving again as subsurface water pressures built up. The Thistle was one of these. On 13 April 1983, it began to move, and within a couple of weeks, a mass of rock and mud amounting to more than 2,000 cubic meters was in motion.

It was one of the largest active slides in the country, one that led finally to losses of $200 million. The slide moved across Spanish Fork Canyon, creating a dam, 60 meters high and over 2,000 meters long. Behind it a lake formed as the Spanish Fork River backed up. The community of Thistle, Denver and Rio Grande Western Railroad and switchyards, and U.S. highways 6, 50, and 89

Figure 5.15. The Wasatch Fault runs through a densely populated part of Salt Lake City and can be seen here as a long, dark line near the middle of the picture. The fault is almost 400 kilometers in length, and it poses the greatest earthquake hazard to the state of Utah because of its proximity to the majority of residents.

were all inundated. One of the disturbing features was the presence of materials that had been involved in previous failures. In fact, as far back as 1967, and again in 1971, this slide had been investigated and listed as dangerous under conditions of heavy rainfall.

Analysis of the slide materials coupled with information from earlier reports indicate that surface and subsurface drainage would have prevented failure and reduced the landslide hazard to one of minor slumping. Estimated cost of these mitigation measures would have been less than 0.1 percent of the costs resulting from the slide. In summary, the data show that the Thistle landslide was recognizable, predictable, and preventable. Some have suggested that mitigation costs for landslides are prohibitive because the problem is so great and there are so many landslide problems all across the nation.

Mitigating Damage

Based on their experience with the Thistle landslide, several engineers suggest that a catalog of all risk locations in the nation can be compiled using aerial photographs and topographic maps. Based on this initial overview, greatest risk areas could be studied in detail by air or on foot. Finally, site analyses could be conducted on places presenting the greatest hazards. The last-named would represent less than 10 percent of the total, and based on the experience of these engineers, average mitigation costs would be on the order of 1 to 100; that is to say, it would cost $1 to prevent $100 worth of destruction.

In the three regions we have studied in Chapters 2, 3, and 4, volcanic eruptions and the associated earthquakes, landslides, and tsunamis dominate public environmental concerns. The U.S. Geological Service is very much aware of this—and hence the major thrust of the 1990s to expand research and, at the same time, launch the Volcano Hazards Program. This is how the latter was defined: "The combination of present observations and the reconstructed history of volcanoes, both currently active and recently active, allow scientists to estimate the types of hazards and the likelihood of their occurrence anywhere in the United States" (*USGS Yearbook*, 1995).

REFERENCES FOR FURTHER STUDY

Allegre, C. *The Behavior of the Earth: Continental and Seafloor Mobility*. Cambridge: Harvard University Press, 1988.

California Geology. Sacramento: California Department of Conservation. Bi-monthly publication.

Decker, R., and B. B. Decker. *Volcanoes*. New York: W. H. Freeman, 1989.

Fisher, R. V., G. Heiken, and J. B. Hulen. *Volcanoes: Crucibles of Change*. Princeton, NJ: Princeton University Press, 1997.

Hansen, R. J., ed. *Seismic Design for Nuclear Power Plants*. Cambridge: MIT Press, 1970.

Jordan, D. S., ed. *The California Earthquake of 1906*. San Francisco: A. M. Robertson, 1907.

Keefer, David K., ed. *The Loma Prieta, California, Earthquake of October 17, 1989– Landslides*. Washington, D.C.: U.S. Geological Survey, 1998.

McPhee, J. *Basin and Range*. New York: Farrar, Straus, & Giroux, 1980.

Rosenfeld, C., and R. Cooke. *Earthfire: The Eruption of Mount St. Helens*. Cambridge: MIT Press, 1982.

USGS Yearbook. Renton, VA: USGS, 1995.

Washington Geology. Olympia: Washington State Department of Natural Resources. Quarterly publication.

Yeats, R., and C. R. Allen. *The Geology of Earthquakes*. New York: W. H. Freeman, 1993.

6

APPALACHIAN OROGEN

The Appalachian Orogen stretches from Maine to Tennessee and includes, in addition to these two states, Vermont, New Hampshire, Massachusetts, Rhode Island, Connecticut, New York, Pennsylvania, West Virginia, and Kentucky. A line from south of Lake Erie southeastward toward Washington, D.C., cuts across all the important terrain of the Appalachian Orogen. The land first rises to form the Allegheny Plateau, the northern end of an extensive series of plateaus. To the southeast of these plateaus is the Appalachian Valley and Ridge Province, a belt of sinuous ridges that curves northward through Virginia and much of Pennsylvania. Here the sedimentary rocks were pushed into tight folds during the last period of mountain building in the Appalachians, the orogeny known as the Alleghenian one.

GEOLOGICAL OVERVIEW

South of the Valley and Ridge Province is the Great Valley, a lowland area created over long periods of time by groundwater and surface water as they dissolved the carbonate bedrock, and still farther south is the Piedmont Province. The basement rocks that make up the Piedmont extend beneath the younger sedimentary layers of the Atlantic Coastal Plain and continue beneath the continental shelf, which is nearly level on much of this coast. During the last ice age the growing mass of ice lowered sea level by about 100 meters, leaving extensive land areas on the continental shelf. Rivers cut deep valleys into this new land, and as sea levels returned to their former state, most of these valleys were filled with sediment (Figure 6.1). Some canyons remain. There is additional information in Chapter 8 on the causes and nature of the ice ages, particularly as they affected the Great Lakes.

Figure 6.1. A twisted and folded stretch of metamorphic rock in Connecticut. The ancient movement of rocks and their modification into these forms under intense heat and pressure deep in the earth are well illustrated here, and they are typical of much of Appalachia.

PENNSYLVANIA

Pennsylvania provides a complete cross section of the Appalachian Orogen, so it is a good case study for introducing the various types of terrain we encounter throughout the region. We will find here the remnants of ancient mountains that were, in their time, higher than today's Rockies. The ridges and valleys that are so often regarded as typical Appalachian scenery are also here. This part of the nation is very old, and rocks that were bent and broken in the very distant past when the North American plate collided with another one are still visible. The last of these intercontinental collisions, the one that was responsible for the present Appalachian mountain system, occurred about 250 million years ago. Pennsylvania will also illustrate for us, in its coal resources, the relationship of geology to minerals.

Beginning at the coastal plain and moving northward across the state, we first reach the higher levels of the Piedmont. The plain is almost negligible in Pennsylvania. The hilly terrain of the Piedmont is quite narrow in this state, and because the streams have cut deeply down into their valleys, transportation across this high ground is very difficult. For this reason, George Washington, when he selected the site for the nation's capital over 200 years ago, ensured that there would be ocean access and also land access north and south, not east and west. There is a lowland basin within the Piedmont, the result of down-faulting a long time ago. The rifting that caused this basin also allowed magma to reach the surface and form low ridges. It was up and down those ridges that the Battle of Gettysburg was fought.

Ridge and Valley is a distinctive belt of long wooded ridges, about 300 meters high, with broad agricultural valleys between them. Sandstone, shale, and limestone are the dominant rock types. Originally they were horizontal layers, but during the Alleghenian Orogeny, they were crushed and uplifted into the forms we see today. Because it was an intercontinental collision that caused the folding, the ridges run parallel to the coast. Many of the valleys have large deposits of anthracite, or hard coal, the outcome of metamorphosis when the uplifting occurred. This is the only economically significant deposit of this kind of coal in the nation. Environmentalists are very unhappy about the use of the strip method to extract the coal. They feel it does permanent damage to the environment.

The largest landform region is the Allegheny Plateau, an area of rolling uplands cut by steep, deep rivers, extending all the way from upper New York State to Alabama. The deeply scored landscape is difficult to cross, and the soils are too rocky for good farming. The heart of the region is around Pittsburgh where rivers such as the Ohio and Monongahela cut through sandstone, shale, and coal seams. Coal is transported on these rivers. To the north, near Lake Erie, there is a glaciated stretch of plateaus, low in relief and well endowed for farming due to the rich glacial deposits. Glacial sediments blanket this area.

LEGACY OF THE ICE AGES

Glacial drift is the term used to describe the different kinds of rock debris left by ice. If a lake or river gets blocked, forming a lake, the material deposited in the lake over thousands of years is sorted into layers of silt, sand, or gravel, with the finest silt sediments near the surface and the heaviest at the bottom. In time, long after the ice and lake disappear, this is the kind of foundation that makes excellent soil, as farmers know so well. *Till* is the term for rock fragments of all sizes and types deposited directly by the ice. Where it is deposited at the front of the ice sheet, it forms, once the ice has melted, a series of irregular heaps, called *moraines*, interspersed with hollows or lakes.

Under the enormous pressure of an ice sheet, melting occurs at the bottom, and streams of meltwater flow out, carrying deposits of coarse alluvium, which are deposited near the ice margin. Thick layers of these outwash sands and gravels are found today on the north shore of Long Island and are mined for construction projects. The ice sheets affected almost all of New York State and we will look at the different effects they had on different terrain (Figure 6.2). The New England states were also covered by that advance, as were parts of the states bordering the Great Lakes. Before finally retreating, about 10,000 years ago, the ice had changed the appearance of the whole landscape and changed the character of the surficial deposits.

Figure 6.2. Landslide and subsidence in a northern area of New England due to seepage and subsequent failure of glacial clays.

Debris from ice activities dammed up rivers, transported millions of cubic kilometers of preexisting soil and rock, and changed the age-old drainage patterns of the state (Figure 6.3). No other geologic process changes the landscape so quickly. As melting took place, huge lakes were formed behind blocked river channels (Figure 6.4). Many of these subsequently drained away, but some remained, including the Great Lakes system.

Construction work has to take special care with foundations. What might have been considered dependable bedrock may well be quite unable to take much weight, and so deep pockets of weathered bedrock were found in metamorphic rock during the planning of New York City's water supply system. These had been deposited by the glacial ice and proved to be poor foundation material.

Ice sheets are powerful eroding agents. They scrape away and grind down the solid bedrock, leaving scratch marks called striations that show, even today, the direction in which the ice moved. One characteristic landform has a smoothly rounded striated surface on the side from which the ice came and an irregular, steep face on the other side where the slowly moving ice plucked out rocks. In some locations of weak bedrock, especially where river valleys ran parallel to the ice movement, the ice scooped out U-shaped troughs. The Finger Lakes of northern New York State were carved out in this way.

The southern shores of Lakes Erie and Ontario have numerous drumlins, thousands of them, and nearby are large numbers of lakes of all sizes. These

Figure 6.3. This photograph shows a 1.5-kilometer-long debris slide on Dorset Mountain in the Green Mountains of southwestern Vermont in 1976. A wide, clear-cut swath was obviously taken through this wooded area, so it must have been a fast-moving slide for that to happen.

are some of the more dramatic illustrations of the ways in which historic drainage patterns were completely disrupted by the ice sheets. I have already mentioned the impact of the ice on the states in the Northeast. There, too, particularly in the north, there are numerous lakes, often accompanied by marshy terrain due to blockage by glacial debris of the older watercourses.

Toward the end of the last ice age the Niagara River was flowing over the escarpment that is named after it. The ancient plunge pool carved by the falling water can still be seen. The top layer, the caprock, of the Niagara Escarpment is made of a strong formation that resists erosion. Directly underneath it is a much softer layer of shale that is easily eroded. Falling water eats away at the shale so that the capstone is undercut and breaks off periodically in large blocks. Over the more than 10,000 years since the ice finally retreated, the falls have cut back upstream about 11 kilometers.

Tributaries of the U-shaped troughs, because they do not lie in the direction of the ice movement, are less affected by the ice. Thus, once the ice has melted, these streams are left high above the steep sides of the U-shaped troughs as hanging valleys. Spectacular waterfalls such as those near Ithaca, New York, appear in these hanging valleys, and over time, as erosion does its work, the waterfalls move back into the tributaries. One indication of the power of ice was found in the ancient Hudson River, which ran parallel to the path of the

Figure 6.4. Glacial lake terrain from northern New York State.

ice. This river valley was scoured to a depth of 240 meters below sea level. Consequently, many tributaries, including the Mohawk River, became hanging valleys, and they now cascade down a series of waterfalls and rapids to the Hudson River.

A large lake, Glacial Lake Albany, filled the Hudson Valley during the last ice age. It was 50 kilometers wide at Schenectady and stretched 320 kilometers from Glens Falls to New York City. At Albany it was 120 meters deep. Its southern end was dammed by a terminal moraine, and its northern extremity was defined by the front of the retreating glacier. Many of the cities and towns of the Hudson Valley are located on deltas built by streams and rivers that flowed into Glacial Lake Albany. These cities include Kingston, Hudson, Albany, Schenectady, Saratoga Springs, and Glens Falls. We will see examples of landslides in or near these places later in this chapter, many of them caused by weaknesses in these glacial deposits.

The Adirondacks are covered with till like other locations, but it is a sandy deposit because of the kind of bedrock (Figure 6.5). The hard metamorphic rock of this region is made of sand-size or larger mineral grains, so glacial grinding produced the sandy deposit type of till. In other places, such as the St. Lawrence–Champlain Lowlands, where the bedrock is shale or limestone, the till deposits are silt and clay. Many of the deposits in the Adirondacks were

Figure 6.5. Whiteface Mt. Olympic Ski Lodge, 1995, in New York State Adirondack Mountains. The area experienced heavy rains at the time. Drainage channels were blocked and overflows saturated the steep slopes, leading to the damaging debris flow shown here. Whiteface owes its name to the numerous landslide scars on its slopes.

made by meltwater from the ice sheets. They take the form of eskers, long, winding narrow ridges along the shores of Adirondack lakes, or else they project into the lake from the shore.

South of the lowlands that border the Great Lakes lies the Allegheny Plateau. Here the dominant action was by the continental ice sheet. The bedrock of the hills was stripped bare, and unsorted till, as much as 3 meters deep in places, was deposited. In the valleys the till was 100 meters or more thick, but the deposits were layered, that is to say, water carried the till and so the heaviest and biggest particles are at the bottom, whereas fine silt is at the top. Some much older deposits are found in valleys that ran at right angles to the main direction of the ice advance. In these valleys we now find glacial till from earlier phases of ice, still undisturbed because they experienced much less ice scouring. Water-deposited material from 30,000 and 50,000 years ago has been identified.

Like New York and other northern states, Connecticut's landscape was fundamentally changed by the glaciers and ice flows of the last ice age. The surface of the land was modified, and all habitats transformed. Virtually all of the state's bedrock was covered with glacial drift from the retreating ice as it gradually melted. Much of the Central Valley is different from other areas because it was blocked by drift for a time in the course of glaciation, allowing a lake to form (Figure 6.6). Over time the sediments deposited in the lake were sorted by the

Figure 6.6. The Connecticut River, in Connecticut, winding its way through the Central Valley. The land here is very flat, a former glacial lake from the Pleistocene ice ages, and because it is a former lake bottom, the soils are rich and free from stones. Flooding is a constant threat, due either to meltwaters from upper reaches of the river or to storm tides from hurricanes.

water. After a few thousand years this lake collapsed, and the water drained away, leaving an extensive flat area of rich soils that today provides excellent farmland.

SURFACE EROSION

Landslides

Landslides are a substantial problem in New York State. On average they cause $10 million worth of damage annually, and over the past century and a half, they have killed more than 70 people. The following is an example of one type of landslide that is directly traceable to the deposits left by Pleistocene glaciation. It occurred in April 1993 on the west side of Tully Valley, approximately 24 kilometers south of Syracuse, and my information about it comes from a report by Robert H. Fickies of the State of New York's Geological Survey Division (Figures 6.7 and 6.8). This valley had been glaciated, and because it lies in a north-south direction, the end appearance was U-shaped. During the melt phase, it was a glacial lake for a time, and in that period, thick layers of a red silty clay were laid down in the bed of the lake.

Figure 6.7. Western edge of the Tully Slide in Onondaga County, New York State, 1993. This happened in a U-shaped valley, sometimes called a "dry finger lake," that had been glaciated in the last ice age and in which a proglacial lake formed. Lacustrine deposits from a lake of this kind are not very stable, and they gave way here when the area experienced very heavy rainfall even though the slope was only 10 degrees.

The slide broke from the western valley slope and moved quickly across the valley flat. More than 50 acres of land were affected, three homes were severely damaged, and four other homes had to be evacuated. Mud to a depth of 4 meters covered a 400-meter stretch of roadway. Several people had to be rescued from their homes by helicopter. The slope failure took place within two major soil units, a lower red clay and an overlying sandy material that thickens toward the west side of the valley. The gently sloping hillside that failed had a slope of only 10 degrees, hardly the kind of terrain where one would expect to see a major landslide; still, it was the largest landslide known to have occurred in New York State within the past 75 years.

The immediate causes of the slide were the heavy rains of April 1993, as much as 19 centimeters and accompanying snowmelt. This amount of saturation weakened the underlying clay and, at the same time, added weight at the western side of the slide area. The soil mass that broke from the hillside was able to slide within a matter of minutes across a water-soaked, clay-rich ground surface. Experts who examined the area a short time after the event felt that the slide mass was marginally stable but that this stability could be threatened by water flowing from the head scarp. Steps have now been taken to minimize this danger. Overall, this slide is typical of the many that occur in the Hudson Valley, all of them related to the lacustrine deposits from the Pleistocene glacial lakes (Figure 6.9).

Figure 6.8. The debris and destruction caused by the Tully Slide in 1993. Fifty-five acres of land were affected, three homes were either destroyed or severely damaged, and people from four other homes were forced to evacuate. Mud, 4 meters deep, covered a local road for a distance of 400 meters.

Landslides can be very small or very large, and they can move at slow or high speeds. They are activated by rainstorms, by earthquakes, or by human activities and usually strike with little warning (Figure 6.10). Some clues of an impending slide can be found in things like doors or windows jamming for the first time, cracks appearing in walls or foundations, underground breaks in utility lines, or tilting of fences and retaining walls (Figure 6.11).

Subsidence

Like landslides, subsidence is an environmental hazard found in many parts of the country. In New York there are areas of karst topography where ground-water flows can eat away the limestone rock, causing collapse of the surface. Sometimes mining can take away so much material in its underground work that a partial or total cave-in occurs. Some events, like the following, lead to more than just a surface collapse or a hole in the ground.

In Livingston County, about 40 kilometers southwest of Rochester, the Retsof Salt Mine has been in operation for more than a hundred years. Its underground workings, at a depth of 300 meters, extend over an area of 6,500 acres. It is the largest salt mine anywhere in North or South America, and it supplies road salt to 14 states in the northeast part of the nation. In March 1994 one room collapsed, and a month later, an adjacent one caved in. Two large circular hollows

Figure 6.9. Lexington Avenue slide, Troy, New York, 1987. Like many other settlements associated with the Hudson River Valley, much of this town was constructed on glacial lake clays, materials that have a record of failing in the regions that were covered by glacial ice during the last ice age. Troy is the location of the earliest recorded landslide in the United States. That was in 1837. Lexington Avenue, at the head of the slide, had to be closed at that time for several months.

measuring 200 and 300 meters in diameter appeared at the surface above the two mine rooms. A road bridge was partly damaged, forcing the closure of a state highway.

Within the 300 meters of overlying rock, two aquifers had been the source of water for a number of local residents who depended on wells. When the ground collapsed, a new cavity for water was created, and both aquifers diverted their underground flows to this new outlet because it provided a quick access to a lower level within the mine. Gradually the flow rates of water into the mine increased until they reached 18,000 gallons per minute, a quantity that almost cut off all supplies to the wells. Furthermore, as underground water flows decreased in the areas of wells, ground compaction took place, and additional subsidence began to appear 5 kilometers southwest of the initial collapse.

Like New York State, Pennsylvania has extensive areas of karst topography, that is to say, land surfaces that have numerous sinkholes as well as other features all caused by groundwater dissolving the carbonate bedrock underground. All of the Appalachian Mountain section of the state south of the last ice age's southern edge has extensive areas of this kind of bedrock, that is, limestone, dolomite, or marble (Figure 6.12). Williamsport is an approximate

Figure 6.10.   Landslide in Albany, New York, 1968, one of hundreds of earth slump or flow slides experienced in New York's Hudson Valley. These slides occur in glacial lake clay deposits whenever they become saturated with water. This particular slide created the 13-meter-high wall at its source, as you see here.

northern boundary for this karst territory. Sinkholes come in all sorts of shapes and dimensions. In these areas of Pennsylvania they are usually circular and range in dimensions from two to seven meters in diameter and the same range in depth.

A comparison between settlement patterns and the distribution of karst topography shows that many of Pennsylvania's larger towns and cities are built on top of carbonate rock. Development in these areas affects runoff because rainwater, instead of being absorbed by vegetation or allowed to infiltrate naturally into the ground, flows quickly across roofs, streets, and parking lots and is then directed into artificial drainage channels. The result is a concentration of water and therefore increased dissolution underground where the drainage channels are located (Figure 6.13).

Buried utility lines in karst terrain are particularly vulnerable, as they are often laid out in trenches with crushed stone or compacted soil underneath. Water collects in these channels because they provide easy paths; so because of the significant volume of water, there is a high risk of dissolution, subsidence, and collapse of the utility line. Some lines are more vulnerable than others. Storm sewers are large-diameter jointed pipes, and water can escape around seams. Water mains carry huge volumes of water under pressure, so it is not

Figure 6.11. Catskill slide, New York, 1996, one of a number of slumps in Catskill over the years. Much of this Hudson Valley city is built on old glacial lake clay deposits. This slide cut High Street and destroyed one home. Clues of imminent failure had been given months before.

unusual for leaks to occur when oxidation takes place on the metallic pipe (Figure 6.14).

A water main collapse in a karst area can be quite spectacular: The pressurized water operates like a power hose. It is not uncommon to see large sections of a highway collapse or buildings get knocked over when this kind of accident happens. The most serious risk of all is the chance of an underground natural gas line being ruptured. Escaping gas can reach the surface quickly, and it only needs a spark to cause an explosion. All of these dangers are due to human activity, interference with age-old underground drainage systems, or negligence regarding utility line maintenance.

States to the north and east share New York's problems from the last ice age. Vermont, New Hampshire, Maine, and Massachusetts all experienced the disruption of ancient drainage channels and debris of all kinds as the ice and glaciers did their work. In addition, even though most of this area is less urbanized than states farther south, toxic chemicals found in water and air alike continue to be a threat to health. Acid rain, the presence of acidic pollutants in rain, is one hazard that is hard to correct in any one state because the problem may originate elsewhere and do damage in quite a different place. The northeastern states lie in the path of prevailing winds, and so they receive the airborne

Figure 6.12. This area was a glacial lake in Pleistocene times, so the lake bed deposits are poorly consolidated. In spring, slides and slumps are common, as can be seen in this scene from Monongalia County, West Virginia.

wastes from the big industrial centers farther west, especially the sulfur dioxide emissions from power plants.

West Virginia and Kentucky have many of the problems we have seen in Pennsylvania because these two states are also major producers of coal. In the Appalachian Region as a whole, the top producer is Pennsylvania, with West Virginia second and Kentucky third. One issue that we will examine shortly— and that is a serious hazard in all three states—is subsidence from coal mining. Permanent reserves of unmined coal protect the central sectors of most of the towns concerned, but commercial development and new residential subdivisions have gone beyond these protected areas. Because of this, every year, dozens of buildings are damaged by mine subsidence, about one third of the destruction coming from abandoned mines.

Flooding

When we think of floods, we automatically associate them with the great Mississippi system because of the huge volumes of water involved and the

Figure 6.13. A steep slope in Morgantown, West Virginia, where weak shales and unconsolidated material are exposed. The result is slope failure, and unless something is done about the problem, the building shown in the photograph will be undermined one day.

enormous damage to lives and property whenever that region is hit with high levels of rain or snowmelt. But flooding is a national hazard, and few areas of the country escape them for long. Connecticut is an eastern coastal state, so it has to cope with flooding from hurricanes as well as from its own main river.

The source of the Connecticut River is close to the Canadian border. It flows south from there, serving as a state boundary for Vermont and New Hampshire, then crossing Massachusetts into Connecticut to reach Long Island Sound at Old Saybrook. The Metacomet Ridge, an upland area of basalt rock, diverts the river from the southern part of the state's central valley. Like other great rivers of the world, the Connecticut deposits a thin coat of silts and muds during its annual flood, greatly enriching the surrounding farmland, New England's best.

At times these normally beneficial floods become destructive when rain is heavy or an unusually thick snowpack melts. The March 1936 flood, when water rose about 11 meters above the normal base level at Hartford, caused $25 million of damage, a big sum for that time. Much of Hartford is built on the floodplain. Other floods, all of them 10 meters or more above base, occurred in 1938, 1955,

Figure 6.14. The house shown in this photograph collapsed overnight into a sinkhole that had been caused by a leaking water main. The large volume of water involved quickly eroded the limestone rock on which the house was built. The site of the sinkhole is western New Jersey at the border with Pennsylvania.

and 1984. The last two or three point to an altogether new hazard, the effects of accelerating suburbanization. As soil gives way to blacktop and roadways, the new smooth surfaces increase the runoff rate, and there are bigger floods. Before 1936, in the more than 300 years during which flood records were kept, flood level never rose as high as 10 meters.

Coastal Erosion

Connecticut and Rhode Island represent the northern extremity of the extensive coastal plain that stretches all the way to the Mexican border. Although in many ways these two states are not typically coastal because of the sheltering influence of Long Island, they do illustrate many of the environmental hazards that affect all the eastern coastal states. Hurricanes are not uncommon, and one effect of these is the raising of freshwater levels by storm surges and hence a heightened risk of flooding in the Connecticut River. The last great hurricane to strike Connecticut was in 1938, but smaller ones hit the region in 1944, 1954, 1955, 1960, and 1976.

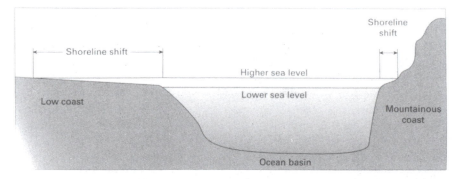

Figure 6.15. The slope of coastal terrain determines the extent of damage when storms, high tides, or river flooding occur. Small vertical changes in water level shift coastlines dramatically on gently sloping coasts but cause only minor shifts on steeper shores.

The low gradients of the rivers as they reach the sea create salt marshes and mud flats all along the coast. The nutritious sediments being dumped provide feeding grounds for many land, air, and water creatures. The most extensive marsh system reaches up into the Connecticut River, gradually changing from salt marshes to freshwater meadows near Hartford. New Haven was founded in a marshy area in 1638, a good site for protection from Indians.

Several processes erode the coastline and transport sediment from place to place. Longshore drifting, the movement of sediment along the coast, occurs when wind drives waves to the shore at an angle. Wind itself also moves sand from place to place as it builds sand dunes, and then the dunes, over time, move along the beach. Tidal currents are yet another force for changing the distribution of beach sediments. Because of the narrow passage between Long Island Sound and the open ocean, the ebb and flow of the tides create strong currents that move sediments from place to place.

The most worrisome geologic hazard facing Connecticut today is the rise of sea level, perhaps due to a global warming trend. The U.S. Environmental Protection Agency predicts that the sea will rise two or more meters in the course of the twenty-first century. If that happens, given the low gradient of the coastal region, tens of thousands of people will lose their homes, and parts of New Haven, Bridgeport, and Hartford may be abandoned. When a coastal area has a very low gradient, then a very small rise in sea level affects a large tract of land (Figure 6.15).

EARTHQUAKES AND OTHER HAZARDS

Most of the world's bigger quakes are found at the margins of the great tectonic plates. New York State is far from any of these margins, so we would not expect to locate much earthquake activity here, and, relatively speaking, that is what we find. California, for instance, has 10 times the amount of New York's earthquake activity. The Pacific coast of Alaska is even more active than California. Nevertheless, over the last 250 years, New York has experienced more than 400 quakes of magnitude 2 or more, mainly where earth movements of one kind or another force rock to move along any one of the state's numerous fault lines (Figures 6.16 and 6.17).

Earthquakes, like landslides or ground collapse, are serious environmental hazards in New York State. There are also the risks of floods and hurricanes. However, the biggest environmental problem facing the state at the present time is water pollution, a human-induced hazard. In many places industrial and agricultural chemicals, radioactive wastes, sewage, and road salt pollute streams, lakes, reservoirs, and groundwater. Sources of these damaging pollutants are numerous—mining and smelting, petroleum production, and disposal of biological and toxic wastes being a few of them.

The first step in dealing with these dangers to human health is geological assessment of both sites of industries on the surface and groundwater conditions, including the sizes of aquifers. The potential damage from particular wastes and acceptable levels of pollutants can then be calculated. Natural areas that can be kept free from commercial and industrial operations enable groundwater to be replenished and thus dilute harmful wastes in the system as a whole.

Bridge Scouring

Connecticut is a good example of another national problem related to waterways though not to floods necessarily. It is channel scouring, the eroding of a river channel around bridge foundations (Figure 6.18). It happens whether or not the river is in flood, and it constitutes the main reason for bridge failures across the country, exceeding in frequency all other types of failure combined. Nationally the cost of scour-related failures is almost as high as the total cost of flood damage on federally-aided highways.

River channels scour and build up sediment in the course of the variety of natural processes that are at work, especially in water channels that have been modified by humans. Bridge crossings are examples of such modifications. They constrict the flow of water in a number of ways. There are about 580,000 bridges in the United States, and 480,000 of these span waterways; so the probability of damage to bridge foundations is high. It is difficult to assess at the design stage the amount of damage that scour will cause to a particular bridge. A great deal of research still remains to be done.

Using the geophysical techniques employed by seismologists to measure the depth of a layer of rock deep in the earth, scientists from the U.S. Geological Service examine the area around bridge footings after a flood. By sending signals from the surface of the water and measuring the strengths and times of their reflection, the types of materials at different depths are identified. This information can be correlated with information already available from other bridge sites and from previous failures of bridges and thus identify danger areas. In-

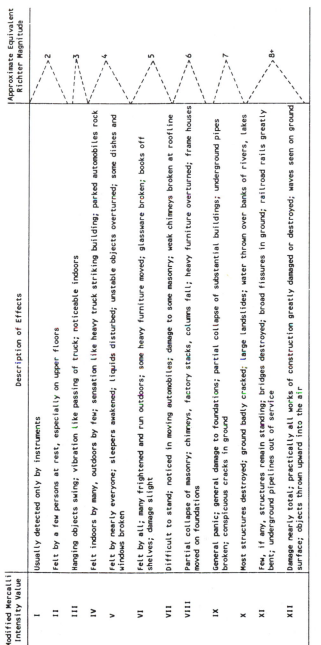

Modified Mercalli Intensity Value	Description of Effects	Approximate Equivalent Richter Magnitude
I	Usually detected only by instruments	
II	Felt by a few persons at rest, especially on upper floors	~2
III	Hanging objects swing; vibration like passing of truck; noticeable indoors	~3
IV	Felt indoors by many, outdoors by few; sensation like heavy truck striking building; parked automobiles rock	~4
V	Felt by nearly everyone; sleepers awakened; liquids disturbed; unstable objects overturned; some dishes and windows broken	
VI	Felt by all; many frightened and run outdoors; some heavy furniture moved; glassware broken; books off shelves; damage slight	~5
VII	Difficult to stand; noticed in moving automobiles; damage to some masonry; weak chimneys broken at roofline	
VIII	Partial collapse of masonry; chimneys, factory stacks, columns fall; heavy furniture overturned; frame houses moved on foundations	~6
IX	General panic; general damage to foundations; partial collapse of substantial buildings; underground pipes broken; conspicuous cracks in ground	~7
X	Most structures destroyed; ground badly cracked; large landslides; water thrown over banks of rivers, lakes	
XI	Few, if any, structures remain standing; bridges destroyed; broad fissures in ground; railroad rails greatly bent; underground pipelines out of service	~8+
XII	Damage nearly total; practically all works of construction greatly damaged or destroyed; waves seen on ground surface; objects thrown upward into the air	

Figure 6.16. Historically the Richter Scale was used to define the strength of an earthquake. Now, with more accurate methods available to measure earth movements, the Modified Mercalli Scale is preferred in the United States. It provides a larger number of categories.

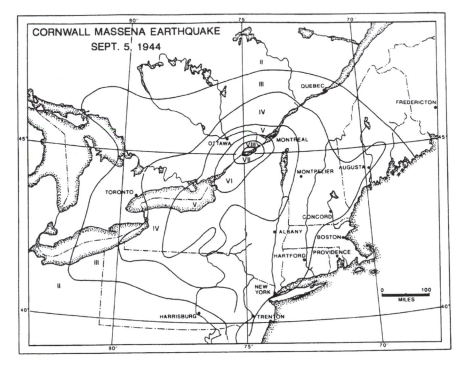

Figure 6.17. The largest earthquake ever recorded in New York State was in Cornwall-Massena in 1944. It had an intensity of 8 on the Modified Mercalli Scale. It was strong enough to damage even well-constructed buildings. While this was a rare occurrence, it must be noted that numerous smaller quakes hit New York State every year.

vestigations of this kind have already been conducted on 56 bridge sites in 14 states.

At the Baldwin Bridge, which crosses the Connecticut near Old Saybrook, close to its point of entry into Long Island Sound, several scour holes were found using the geophysical techniques. At one pier, a seven-meter-deep hole developed without any significant flood event having occurred within three years of the start of construction. Farther upstream, at the Bulkeley Bridge near Hartford, scour holes were detected at four out of eight piers.

In pursuit of a more detailed data bank, extensive research was carried out on a 1995 bridge failure in California where flood conditions collapsed a bridge and caused the death of seven people. Stream data for the main river as well as its tributaries was collected for both upstream and downstream stretches. High-water marks were noted and geomorphic conditions examined. Quantitative observations were made on streambed lowering, bank erosion on both sides, amounts of sediment on the riverbed, and possibilities of wood debris. The final outcome was an estimate of the overall conditions that were present at the time of the failure.

Figure 6.18. New York State bridge over Scholarie Creek in 1987 after storm-induced flooding caused severe erosion of the bridge's footings and surrounding soil. The collapse of the bridge was sudden, and five motorists were killed when it happened. The bridge was subsequently rebuilt with much deeper footings.

RADON

Radon is a natural colorless and odorless radioactive element found almost everywhere, and as long as the quantities ingested or breathed are small, the danger to health is no more serious than it would be for all the other kinds of environmental radiation—sun's rays, medical x-rays, and nuclear power plants. Radon and its progeny are decay products of one type of uranium, and the decay process is accompanied by the emission of alpha and beta radiation, which have health risks only if the quantities are substantial. It has a short half-life—3.82 days.

Most of the radon produced in the earth's crust remains in the soil or rock. A small amount is released into soil gas and groundwater. The gas can thus be carried into the outdoor atmosphere, indoors, or into the groundwater. Large numbers of people in Connecticut draw their water supplies from groundwater, so there is some concern over the risks to which people are subjected. There is still a good deal of uncertainty about what constitutes a truly safe level of radon, but—as we will see—federal authorities have decided to establish one. The Connecticut Academy of Science and Engineering was one of the first to tackle this problem of standards. Its 1998 report "Radon in Connecticut" is a landmark document on the subject. It tabulates amounts in wells and surface water and suggests courses of action that will ensure safety.

Radon is a negligible risk in the outdoors, but it is a different story when the gas concentrates in the soil beneath a residence and there is little exchange of air between indoors and outdoors. Such an arrangement adds significantly to the amounts breathed. Additionally, gas is released within the home from drinking water, especially if it happens to come from groundwater, and can then be either breathed or ingested. Measurements taken by the Connecticut Academy of Science and Engineering show that people in the state who draw their water from wells or groundwater have a higher risk than those who use surface water.

Federal and state agencies have been investigating the radon hazard because one in every seven lung cancer deaths in the United States could be a result of radon. It is as serious as that. It is also regarded as a causative factor in leukemia. Oregon's Public Health Division studied the relationship of radon to the geology of the urban region in and around Portland. It concluded that it was low, but when indoor radon values were collected, some parts of Portland had very high values. The Environmental Protection Agency has chosen a value of 4 picocuries per liter of air as the maximum beyond which action to reduce radon levels is strongly recommended. This figure of 4 represents the equivalent of receiving 200 chest x-rays in a year.

Nevada conducted statewide tests of radon in 1990 and 1991. Among the 2,000 homes inspected, 10 percent were above the critical level of 4, and one was as high as 47. The individual average for the state was approximately 3, and as we would expect, the highest readings came in wintertime when doors are closed. There is a close link between radon levels and local bedrock. Because granite can contain above-average amounts of uranium, homes close to this type of rock have higher values. Some shales and metamorphic rocks also have significant amounts of uranium.

In all, 34 states participated in the Environmental Protection Agency's radon survey of the early 1990s, and approximately 50,000 homes were involved. About one fifth of these homes exceeded level 4, and the overall average was 3. Some homes had readings as high as several thousand picocuries. It is hard to imagine conditions of this kind. We can only hope that they took urgent action when the levels became known. The federal Environmental Protection Agency points out that if you are exposed to more than 15 picocuries of radon over your lifetime, the result is equivalent to being a two-pack-a-day cigarette smoker.

Several hundred years ago, mine workers in eastern Germany often developed a fatal disease that was given the name "mountain sickness." Only in recent times was this disease recognized as primary lung cancer, with the source of the trouble being radon from uranium in the mine. Levels of radon and its progeny are very high in this part of Germany. A good deal of research remains to be done on the dangers of lung cancer from radon and its progeny, but the various moves that have been made have at least made it clear that this is a serious environmental hazard.

COAL

Coal is a sedimentary rock, formed in the ancient past, about 300 million years ago, and today is found in many parts of the United States, including Pennsylvania. Because of its historic association with this state, it serves well to illustrate the relationship between geology and mineral resources. Coal is different from other rocks because it does not come from minerals but from plant remains. About 300 years ago in the Pennsylvania period, as that time was called, western Pennsylvania was close to the equator and had a warm, moist, tropical climate. Vegetation was abundant, and as ferns, trees, and leaves fell into the swamps and were prevented from oxidation by the water, they gradually accumulated into thick masses of peat.

Subsequently the peat was subjected to heat and pressure by overlying sediments, and the more easily vaporized compounds were gradually forced out. This process concentrated the carbon and eventually turned the peat into coal. As the pressure increased, the more volatile compounds were forced out, and the coal thus became richer and richer in carbon, the main source of heat in coal. Most of Pennsylvania's coal is soft, or bituminous, and is mined in the Appalachian Plateaus Province. To the east in the Appalachian Mountain Region, where vegetation was subjected to deeper burial and higher temperatures, a much harder coal, anthracite, is mined.

Most of the coal that is mined contains all kinds of additional matter such as volcanic ash, sandstone, and small quantities of every kind of metallic and nonmetallic elements. These are the things that were carried into the original marshes by wind and water, and they affect the value of the coal. The higher the percentage of carbon in a given amount of coal, the greater will be its heat content (Figures 6.19 and 6.20). About 85 percent of all coal is used in the production of electric power, while the rest is mostly devoted to steelmaking.

From 1880 to 1950 coal was the nation's primary fuel for industry and transportation. It still is a major source of energy and is available in huge quantities. It is also widespread, underlying about 13 percent of the nation's land area. Almost a billion metric tons of coal are produced each year, 85 percent of which is used for the generation of electric power. Reserves are enormous. At the present rate of consumption, they could last a thousand years. Most of the coal now being consumed is bituminous.

Coal from Pennsylvania's Plateaus Province is low in sulfur content and therefore attractive from an environmental standpoint. In addition, the coal seams are relatively flat, an important consideration in mining costs. The lower percentage of sulfur is a welcome feature, but there are other and very big questions being asked about the deleterious effects of this energy source. These

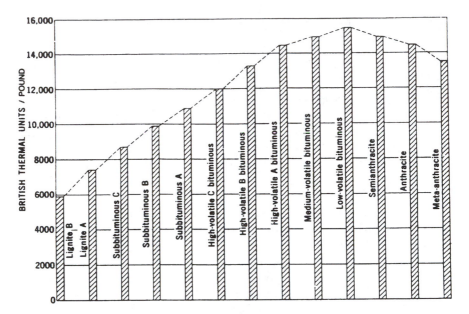

Figure 6.19. Heat values of different types of coal, an important consideration because most American coal is used to generate electric power. At one end, lignite has very low heat quality, whereas at the other end, the different types of anthracite are lower in heat than some bituminous varieties even though anthracite has higher carbon content.

center on pollution from coal combustion and damage to the physical landscape, particularly where there is mass removal of topsoil by dynamite to make it easier to reach the coal.

The wide range of environmental problems being raised includes the following: The physical environment is seriously disturbed; coal transport is expensive and adversely affects communities near rail lines; utilities, even after flue gases have been scrubbed to remove sulfur and other obnoxious components, still remain the main contributor of carbon dioxide and sulfur dioxide emissions into the atmosphere; and mine sites contain harmful chemical elements that are picked up by water and contaminate surface streams and groundwater.

Fears of global warming are now added to these, and new demands are being made to reduce drastically the amounts of carbon dioxide being released into the atmosphere. In the early months of 2000 in West Virginia, the use of dynamite to remove overburden and strip mining for the coal, one of the latest phases of mining technology, led to widespread protests. We are likely to see many more. All these are challenges that affect the livelihoods of coal-mining communities. A look back at earlier mining methods and developments over the twentieth century will give us a better understanding of these environmental issues.

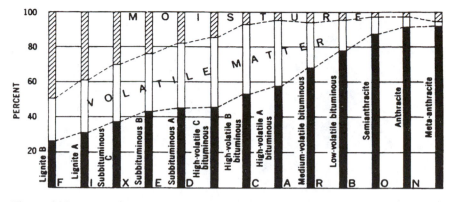

Figure 6.20. Large quantities of volatile material are not desirable in coal, either for heating purposes or for use in blast furnaces. The low-volatile bituminous coal is well suited to production of coke, so it is popular in the manufacture of iron as well as in electrical generation.

Most coal mining is underground, and traditionally this was done by the room and pillar method; masses of coal are removed, creating rooms, while equally large masses are left standing as supports for the overlying rock and soil. This was not a very efficient way of extracting coal, but quantities have always been, and still are, so large that 50 percent being mined was an acceptable level. With increasing demand for coal from late in the nineteenth century right up to the middle of the twentieth, these earlier extravagant methods gave way to more and more efficient ones. The West Virginian example I just described is the latest.

Steps to increase the percentage of coal extracted underground from a given seam led to leaving less and less overburden, so that as soon as the seam of coal was exhausted, the surface was allowed to collapse into the mine. This was quite acceptable if the site was in a remote area and no one objected to a depression in the ground, or if fill of some kind was added to level the surface. Unfortunately, not every mine was cared for in this way. Some were left in their last working state. Years later, sometimes after 50 years, the ground finally collapsed. By that time, urban expansion had encroached on the site, and no one knew that a mine had been there. Buildings suffered in the collapse as they had in karst territory (Figures 6.21 and 6.22).

As technological improvements came into use, higher and higher percentages of available coal were taken out. Pillars were removed in a systematic way once the rooms had been emptied, and an orderly collapse of the mine took place. In a very large number of mines, unfortunately, pillars were left in place and the mine abandoned. Frequently its location was not accurately recorded for posterity, and so, as I indicated, unexpected ground collapse occurred at later times. In Pennsylvania, the most seriously threatened 150,000 acres of land are poten-

Figure 6.21. A grocery store in a residential area of Pittsburgh, Pennsylvania, subsiding because of its location over an abandoned coal mine. It is a common story in this coal-rich part of the country to have coal deposits in particular locations run out and mines then abandoned. In time the overlying rock weakens and sinks.

tially subject to subsidence. West Virginia is a close second. Nationally, coal mining has taken place under 8 million acres of land, and one quarter of that area was subject to some degree of subsidence.

Underground mining is a dangerous occupation. We have looked only at the dangers to those on the surface. Those who work underground know the risks. Death rates have been steadily falling, but they are still far too high. In the 1990s more miners were dying than in any one of the following occupations: agriculture, construction, manufacturing, and transportation. Deaths can be caused by coal gas explosions, roof collapse, or fires when coal dust or methane gas ignites. One underground site in central Pennsylvania had an underground fire burning for 25 years in some abandoned mines.

Safer coal mines are now top priority in this industry. Facing, as it does, major challenges because of the environmental destruction it causes, the industry is taking extraordinary steps to minimize deaths and make sure that at least this aspect of its work will not receive public condemnation. The reality of plate tectonics alerted mining companies to the constant movement of the earth any-where and everywhere. Because these movements are very small, they were ignored in earlier times. The new monitoring techniques use microseismics, lo-cating underground rock noise, and resistance to electrical current flow in rock above and below the mine. Both of these measures indicate increase of stress,

Figure 6.22. Aerial photograph near Sheridan, Wyoming, showing sinkholes above abandoned coal mines. Some of these mines date back as far as the beginning of the nineteenth century. This type of subsidence can be prevented either by leaving enough coal in the seam to support the roof or by adding fill material.

the sort of thing that would be a prelude to rock failure. Roof rock bolts and other supports are immediately installed when warning signs are received.

When an accident happens deep down in the earth where miners are at work, there is an inevitable delay in getting assistance down to the accident scene. If the event is a fire or explosion that suddenly fills the work area with smoke and gas, fresh air is needed immediately to keep the miners alive until help arrives. In the past a situation of this kind meant death. A device known as the SCSR, which stands for self-contained self-rescuer, is now given to every coal miner. It provides a minimum of 60 minutes of breathable air. The SCSR weighs only 2.5 kilograms, small enough to be worn easily on the body, and is tough enough to survive daily usage in the mine.

The small spot of light from the miner's cap used to be all that was available to illuminate the work area. All around, outside that small beam, everything was dark. Now there is a general requirement that all equipment in use at the coal face have an illumination system that lights up the whole of the work area. It is there that most accidents happen. This area of work is increasingly complex because of the new types of machines in use. They are more powerful and more complex, all the more reason for better lighting. Furthermore, because they are more powerful, the cutting machines carry protective canopies to protect miners from flying blocks of coal.

The first coal mines worked as near to the surface as possible and in locations best suited to transportation systems. Once these sites were exhausted, some mines had to go deeper, and that change brought with it a greater risk of explosions because the content of methane emissions from coal is greater at the deeper levels. The answer to this challenge is better ventilation, and up to a point, this has been given high priority in the deeper mines. Where the problem persists—that is, where normal ventilation methods are insufficient—methane drainage techniques were developed. Both in advance of new mining sites and in existing ones, site-specific drainage techniques are now able to secure safe working environments.

There is one more hazard in underground mining. It has already been mentioned in relation to subsidence in Livingstone County where groundwater was profoundly affected by changes in ancient drainage patterns. Every mine changes the flow channels of groundwater and, to a lesser extent, of surface drainage. While the mine is in operation, any increase in amounts of water will be contained and diverted, but after it is abandoned, ease of access will allow groundwater to take the line of least resistance, at the floor of the mine, just as happened in Livingstone County, thus depriving people in surrounding communities of their groundwater well supplies.

Mitigating Damage

Radon can be measured easily with an inexpensive charcoal canister, which can be purchased in most communities and will test for radon in a home over a period of one week. The test instrument must be placed near a likely source such as the basement of a home. The fact that radon is present at a high level in a home nearby does not mean that other homes are also in danger. Local conditions may vary greatly from home to home. If high levels are detected in your home, then action such as better ventilation, sealing of possible entry points at the floor or basement, and adding air fans are typical countermeasures to ensure that the radon is thoroughly dispersed.

REFERENCES FOR FURTHER STUDY

Anderson, B. C., and H. W. Borns. *The Ice Age World*. Oslo: Scandinavian University Press, 1994.

Bell, F. G. *Site Investigations in Areas of Mining Subsidence*. London: Newnes-Butterworths, 1975.

Ebel, J. E., and A. L. Kafka. *Earthquake Activity in the Northeastern United States*. Boulder, CO: Geological Society of America, 1991.

Jensen, D. E. *Minerals of New York State*. Rochester, NY: Ward Press, 1978.

Jorgensen, Neil. *A Guide to New England's Landscape*. Chester, CT: Globe Pequot Press, 1977.

Kates, R. W. *Industrial Flood Losses: Damage Estimation in the Lehigh Valley*. Chicago: University of Chicago Press, 1965.

Keystone Coal Industry Manual. New York: McGraw-Hill, 1981.

McCullough, D. G. *The Johnstown Flood.* New York: Simon and Schuster, 1968.

New York State Museum. *Geology of New York.* Albany: University of the State of New York, 1991.

Stolwijk, J.A.J., ed. *Radon in Connecticut.* Hartford: Connecticut Academy of Science and Engineering, 1998.

Van Dusen, Albert E. *Connecticut.* New York: Random House, 1961.

7

COASTAL PLAINS

The Coastal Plains region extends all the way from New Jersey to the shores of Texas and includes, in addition to these two states, some or all of Delaware, Maryland, Virginia, North and South Carolina, Georgia, Florida, Alabama, Mississippi, and Lousiana. The Plains are a very gently sloping land surface at the eastern and southern edges of the United States, and the submerged continental shelf is a continuation of the same surface. The average breadth of the two together is about 300 kilometers for most of the region. The sedimentary rocks thicken as we move farther and farther away from land. Furthermore, some deposits at the outer edge of the shelf are very old. These two things are evidence that the eastern edge of North America is slowly sinking.

GEOLOGICAL OVERVIEW

The Atlantic and Gulf Coastal Plains form a relatively flat region, dipping slowly toward the ocean with a slope of only 50 centimeters per kilometer. Beneath the surface are varying depths of unconsolidated sediments ranging in age from 100 million years ago right up to the time of the last ice age. For the most part they are composed of sand, silt, and clay, and as they dip toward the sea, their depths increase from less than 100 meters at the inner margin of the Coastal Plains to thousands of meters on the continental shelf. These sediments come from both marine and land sources. Coarser-grained materials from the upland areas are deposited first and finer sediments last, making the areas closest to the sea the least permeable.

The various changes in sea levels over millions of years add their own modifications to the coastal terrain. Groundwater and surface water flows tend to follow the dip of the land, as we would expect. This leads to serious flooding in coastal areas whenever a hurricane strikes and rainfall is heavy, as North Carolina discovered in 1999. In sharp contrast to Pacific shores, the Atlantic continental shelf is wide and gently sloping. In part this is due to the ice ages, which affected the eastern United States more than the west. During periods of low sea levels the shelf was land and was shaped by erosion, just like inland areas today. Many of these features are still evident.

Fifteen thousand years ago, sea level was 100 meters lower than today because the weight of ice had depressed the land. With sea level so much lower, the land surface ran out close to the edge of today's shelf, and rivers flowed all the way to this distant shoreline. Once the sea level rose again, the various river channels were partly filled with sediment, but shelf valleys remained and they are still visible today. Old barrier islands from these ice ages also remain as a series of ridges running parallel with the shore. The Gulf Coastal Plain, while similar to the eastern areas in many ways, is very much bigger in extent, and this feature is well illustrated in the state of Mississippi.

MISSISSIPPI

The Mississippi floodplain forms a large part of this state. It is sometimes referred to as the Yazoo Basin because almost all of that river's delta and floodplain are contained within the larger alluvial lands of the Mississippi, a huge floodplain covering thousands of square kilometers. This large area is almost completely flat. Slight elevations occur along stream banks or on old meander stretches of abandoned streams. Meanders are a universal feature of both the Mississippi and other rivers because of very low elevations and unconsolidated surface deposits and therefore the ease with which an overflow of water can carve out a new channel.

On the banks of the Mississippi, the heavier silts and sands accompanying overflows are the first to be deposited, so they accumulate close to the river, forming natural levees. From there lighter deposits are carried eastward to a zone of backswamps. The endless sequence of river floods requires some control if there is to be any settlement or farming on this highly fertile land. However, as we so often see, whenever there is interference with natural processes, there is a price to pay for our interventions. The natural levees are built higher to force the river to carry greater quantities of water and so reduce the frequency of floods. Thus silt and sand that ordinarily would have left the river during flooding now remain in the riverbed, raising the level of the river and compelling authorities to build still higher levees.

The same problem is associated with dams that are built for flood control. Silt and sand once again accumulate in the riverbed, this time more rapidly because they are being left in the lakes behind the dams. In time, unless these lakes are dredged, riverbed deposits will thicken, the lakes will get shallower, and before long, overflow will begin. There is another problem caused by dam construction, the reduction of deposits at the mouth of the river where they are vital for the life of wetlands. We will examine more aspects of this when we consider the Gulf wetlands. Beyond the edge of the Mississippi Flood Plain, immediately east of the backswamps, there is an area of loess deposits stretching all the way from the Tennessee border to the Gulf. These deposits were taken here by wind from glacial outwash material of the Pleistocene age.

Environmental Issues

In central Mississippi, expansive soils are a constant threat to the stability of buildings and roads. They are found all over the Yazoo River delta, and in one section, appropriately called Yazoo Clay, the sediments are 130 meters thick. The damage caused by these soils is greatest under slopes where there is a downward movement of soil. Subsidence is not a serious problem in the state as a whole, perhaps because there is not a shortage of water, the reason for problems of this kind in California. It is feared that the problem could appear very quickly if groundwater withdrawals increased because such water would come from unconsolidated sediments, the kind that soon collapse under these circumstances.

Flooding is always a major concern in this state. We only have to look at the very low elevations on either side of the Mississippi, and keep in mind the huge volume swings in this river, to know that everyone in the floodplain is under constant threat. It is a difficult danger to tackle because if dams are built on the river to prevent floods, wetland areas suffer and the sea encroaches farther onto the land. In 1999, a controversial plan was proposed to build a bypass floodway in order to protect agricultural land. Intense debate followed as wetland pre-servers were pitted against agriculturalists.

Flooding and expansive soils are not the only environmental hazards facing Mississippi. While the state receives many fewer hurricanes than places on the east coast, the threat from this source is very real. In August 1969, Hurricane Camille made landfall on Mississippi's coast. It was a storm that ranks among the most intense and damaging ever to hit the United States. Earthquakes are also a continuing risk. Ever since the causes and past frequencies of the early nineteenth-century New Madrid quakes were identified, all the states bordering the locations of these earthquakes have adopted a new alertness to danger. The northern border of Mississippi is less that 20 kilometers from New Madrid.

SURFACE EROSION

Landslides

It would be hard to find a major region in the United States where landslide problems are not encountered. It is a ubiquitous hazard, and residents of the Coastal Plains are well acquainted with them. Madison County, Virginia, ex-perienced hundreds of landslides following a severe storm in 1995. Many of the slides became debris flows, masses of rocks, water, soil, and vegetation rushing downslope, destroying buildings, roads, crops, and livestock. There had been previous catastrophic floods in 1937, 1942, and 1972, but none of them caused the 1995 level of destruction.

Geology explains much of this tragedy. Long exposure of bedrock through thousands of years of subaerial erosion produced soils that break easily. These materials are 10 meters deep in places, especially at the bases of steep slopes, and stream erosion moved much of it onto the floodplains of local rivers. A large number of people live on or near these floodplains and are engaged in mixed farming. Quite apart from the condition of the ground surface, the volume and location of rainfall are the other factors that led to the destructive flooding.

The heaviest rainfall came over a 16-hour period with a concentration in one part of that time when rain was falling at 10 centimeters an hour. This is a rare condition and only happens when a thunderstorm stalls over one location. In 1972 a similar thing happened in South Dakota, causing the Rapid City flood, and also in Colorado in 1976, leading to the Big Thompson River flood. Some drainage channels were altered as a result of the flooding and debris flows, so there is a greater risk of flooding in the immediate future until corrective measures are taken.

Rare though the events of Madison County were, the things that happened could very easily be replicated in the central and southern parts of the Appalachians. It is possible to plan for such hazards, and Madison's experience can point the way to avoid a repeat of the tragedy. For example, debris flows caused the worst damage just beyond the mouths of channels. It would be a good preventive measure to avoid building structures at these places and at neighboring channels that are only occasionally filled with water.

Debris Slides and Flows

In the high ground of the southern Appalachians, especially on the Blue Ridge lands of western North Carolina close to the border with Tennessee, on the western edge of the Coastal Plains, there are frequent intense rainfall events that trigger debris slides and flows. Studies of the area show that these earth movements occur over much of the southern Appalachian Plateau, Ridge and Valley, and Piedmont, but the Blue Ridge records the biggest concentration. It is a region of complex ridge and ravine landscapes that rest on highly deformed igneous and modified sedimentary rock units. Soils vary in depth from one meter on the uplands to great depths in some of the valleys.

There are no distinctive preconditions giving rise to these debris slides. They occur when the ground is dry and when it is sodden. Part of the difficulty in correlating slides or flows with rainfall stems from the spotty nature of the precipitation cells and the short time involved in particular intensities. The degree of slope is always important for predicting events, and in the Blue Ridge area, the average is 40 percent, quite a high figure. Many places in California become concerned when the slope exceeds 15 percent, and usually special geologic studies are made when slopes exceed that figure.

The great volume as well as the rock and timber components of typical flows in the Blue Ridge area make them extremely destructive. The preferred paths downslope frequently cross access routes through which rescue personnel and emergency supplies would have to pass in the case of an emergency. As the region continues to experience explosions in tourist volume, the potential for loss of human life grows. The following description of a recent rock slide in this area will help to illustrate some of the above.

In July 1997 a large rock slide hit North Carolina's Interstate 40 close to the border with Tennessee. Several hundred cubic meters of large rock, soil, mud, and water were involved. Several vehicles were caught in the slide, and three injured people required hospitalization. The slope continued to fail throughout the day, so this stretch of highway through the Pigeon River's gorge was closed and remained out of service for about five months. This event was not a complete surprise. Both when Interstate 40 first went through the 30-kilometer section of the Pigeon River's gorge and subsequently when corrective measures were taken, there were continuing concerns over the stability of the adjacent slopes.

During construction in the 1960s several slides occurred, and once the road was opened, rock fall was an almost daily event, in large part due to having had to make cuts as high as 100 meters and 300 meters in length. In 1978 a slide carrying about 300,000 cubic meters of material blocked the highway, and it took nine months to clear and repair the road. A slope repair project started in the 1980s. The roadway was moved a short distance sideways in order to provide a catchment area at the foot of the slope, unstable material in the cuts were excavated, about 13,000 meters of rock bolts were installed in the slope, and wire mesh and horizontal drains were added.

For a time, from 1986 to 1997, things were much better. Slides were less frequent and of less magnitude than before. But then came the 1997 slide, and new efforts had to be made to stabilize the slopes. In addition to the removal of debris and the preventive measures introduced in the 1980s, three separate chain link nets were hung on six-centimeter cables across the slopes. Their purpose was to absorb the energy of rolling rocks and direct them into the catchment area at grade.

Sinkholes

Sinkholes, as we have already seen, are landscape features in many parts of the nation, and we will see further evidence of them as we go on (Figure 7.1). In North Carolina the concentrations are in the southeast part of the state, in such counties as Brunswick, New Hanover, and Pender. The bedrock here,

Figure 7.1. A sinkhole in Shelby County, Alabama. Soluble limestone and other carbonates are present in almost half of the state, with the result that thousands of sinkholes have appeared, causing problems for all kinds of construction including highways.

formed 50 or more million years ago, is a sandy shell-limestone, sponge-like, and probably laid down in an environment of reefs. The appearance of present-day sinkholes is tied to the drainage patterns.

As rivers cut back into the land, the increased gradient and faster flow of water flushed out the sandy and silty infilling of the cavernous limestone. In this process the groundwater was lowered and hence the weight of the over-burden effectively increased. Eventually the sponge-like interior collapsed, and the land subsided. Thus the threefold conditions for sinkhole development here are: (1) a limestone layer with the randomly arranged structure of a reef with sand or clay filling pipes and crevasses and with a soil overburden; (2) a gradient sufficient to wash out the clay and sand interstices; and (3) an outlet for the water and its flushings (Figure 7.2).

Recognition of the danger of sinkholes in this part of North Carolina is quite recent. The geological literature of the 1960s almost ignores them. This is not to say that there was no awareness of ground depressions and numerous sinks, but because of their locations in remote wooded areas and the fact that they did not present problems, they were regarded as curiosities and often used as dumps. Only when new transportation routes and costly military installations were being built was serious attention paid to the inherent risks.

In Pender County, grading for an extension of I-40 coincided with the opening

Figure 7.2. A large sinkhole, locally known as the "December Giant," which formed overnight in rural Shelby County, Alabama, in 1972. It measures 100 meters in diameter and over 30 meters in depth. In cases like this, the carbonate rock underground is gradually eroded by groundwater until one day the surface soil and rock collapse suddenly.

of a Martin Marietta limestone quarry in the summer of 1983. This territory has extensive underground deposits of shell-limestone. Water levels in the quarry were soon lowered to a depth of about 10 meters, and—as I have already noted—this is exactly the kind of development that will trigger subsidence. The groundwater gradient is increased and sand and silt are removed from the limestone interstices below. Within a couple of months the first signs of subsidence appeared in the highway. Following a heavy rainstorm a 7-meter diameter depression formed in the southbound lane. Others soon followed.

To minimize construction delays, sinkhole locations were mapped and organic material removed. Granular material was brought in to backfill the cleaned depressions and bring them up to ground level. Concrete weights were then used to compact both the granular material and any potential underground voids. More granular material was added to restore everything to ground level. These measures have proved sufficient to deter sinkholes over much of the area, but they do not constitute a permanent solution. It needs to be added that solution of the limestone, the main cause of sinkhole development in other parts of the

Figure 7.3. Desperation to protect this hotel near Fort Lauderdale, Florida, is evident from the picture. The building is situated on soft sand, so, unfortunately, neither the sandbags nor the breakwater will suffice if a major storm comes.

country, has not been a significant factor in southeastern North Carolina. Here the pace of limestone solution is much slower than groundwater erosion.

Flooding

The more spectacular accounts of flooding are usually associated with the Missouri-Mississippi system because it is in that basin that the nation's biggest flows of water occur. In the 1990s, however, because of an increase in the severity of hurricanes, the eastern coasts of the country experienced unprecedented flooding (Figure 7.3), and North Carolina was one of the hardest hit in the year 1999. That was the year of Hurricane Floyd, one of the most powerful storms ever to threaten the United States. First warnings came in mid-September when 1 million people were told to evacuate the southeast coast. At that time Floyd packed wind speeds of 250 kilometers an hour, very close to the top category for hurricanes, number 5, and the storm as a whole had a diameter of 1,200 kilometers.

Within a few days it became clear that Floyd was not going to hit land, so the fear of widespread devastation receded, but as the storm moved northward storm surges and heavy rainfall swept beaches and inundated low-lying areas everywhere (Figures 7.4 and 7.5). When it reached North Carolina, it dumped 50 centimeters of rain in less than a day. Every river crested beyond its flood

Figure 7.4. Sinkhole in Winter Park, Florida, 1981.

level, leaving 48 people dead, washing out homes and office buildings, and stranding thousands. Flood damage was everywhere: 30,000 hogs and 2.5 million turkeys were drowned; the Federal Emergency Management Agency fielded requests for housing assistance from 46,000 victims; more than 80 coffins floated away from the Edgemont County cemeteries.

For more than a month, flood conditions persisted. Weather conditions were bad, and rainfall, though less than the levels brought by Floyd, remained high. Over one 24-hour period, an additional 20 centimeters fell. Near Raleigh the swollen Tar River was 2.5 meters above flood stage at one point. Then, in the third week of October, just as river levels began to drop, Hurricane Irene struck, leaving 30 centimeters of rain on all the places that were still trying to recover from the deluge of the previous weeks.

A 100-Year Flood

In July 1994, parts of Alabama, Florida, and Georgia, especially Georgia, were devastated by floods from Tropical Storm Alberto. It was a storm that took 33 lives, and damage amounted to $1 billion. Seventy-eight counties in the three states were declared federal flood disaster areas. Numerous streams crested at levels higher than had ever been experienced before. Highway traffic was disrupted as bridges and roadways were either flooded or washed out.

Alberto grew from a tropical depression that formed off the western coast of Cuba and first came ashore on the Florida Panhandle. Once ashore, it lost energy

Figure 7.5. Wave-cut scarp with exposed Miami limestone, Miami, Florida. This rock face was formed in Pleistocene times during a period when sea level was higher than today.

and was downgraded to a tropical depression in less than a day. It drifted north to a point west of Atlanta, then moved slowly in a southwesterly direction. This slow movement combined with abundant tropical moisture produced very high levels of rainfall. Storm totals greater than 30 centimeters were common in all three states. High rainfall totals of 70 centimeters over a four-day period and 50 centimeters within 24 hours were recorded at Americus, near Columbus in central Georgia.

Again and again, rainfall totals and flood levels in Georgia gave clear evidence that this was a 100-year event, that is to say, a flood that would on average occur only once in a hundred years but one that has a 1 percent probability of happening in any year. The Americus 24-hour rainfall was 2.5 times the estimated 100-year amount, and water levels on some rivers were also more than twice the levels anticipated for a 100-year event. At many locations, gaging stations were destroyed by floodwaters, so immediately following the flood, field crews were dispatched to flag high-water marks along major rivers.

HURRICANES

Cyclonic storms are centers of low pressure with inward-spiraling winds that form where warm and cold air masses meet. In the Pacific Northwest, as we have seen, concerns about these storms focus mainly on the amounts of precipitation they bring. When the warmer air within the storm rises over the colder, there is rain or snow; and if the quantities are very big, there may be damage to the environment. It is a very different story in lower latitudes where temperatures are so much higher and the cyclones consequently much more destructive.

Hurricanes develop over water about 10 degrees north or south of the equator in a manner similar to cyclonic storms elsewhere in the world, then move westward through the trade wind belt. Sea surface temperatures in these latitudes are often at 27 degrees Celsius, enough to stimulate high-speed inward-spiraling winds of 100 to 200 kilometers per hour. The shape of the storm is circular, with a diameter averaging 150 to 500 kilometers. As the hurricane moves toward the eastern seaboard of the United States, or into the Gulf of Mexico, the National Weather Service begins its warning and forecasting activities. As the storm approaches land, the biggest concerns relate as much to the expected rainfall as they do to any physical damage.

Attempts were made in the 1960s to reduce the severity of hurricanes by cloud seeding. The theory was that by seeding the clouds closest to the point of lowest pressure, latent heat would be released, raising the temperature and therefore decreasing the pressure. As a result the pressure gradient is reduced and the maximum wind speeds are also reduced because energy is redistributed around the storm center. Four of these storms were seeded on eight different days, and results showed a 30 percent drop in wind speed on four of these days.

Subsequently the experiment was dropped because additional research revealed that hurricane clouds carry a large volume of natural ice! Reliance is now placed on accurate forecasting and protective measures on shore. The Saffir-Simpson Scale of hurricane intensities was introduced in 1974 to indicate the storm's power. Category 5 is the most powerful, with speeds of more than 249 kilometers per hour, and it is rare that one of such strength reaches the U.S. mainland. Category 1 has speeds of 120 to 152 kilometers per hour.

Just as happened in the case of earthquakes where the Mercalli Scale was introduced, so with hurricanes there is increasing demand for an alternative to the Saffir-Simpson Scale. This measure is fine for identifying the absolute strength of a hurricane in the open ocean but not so good for indicating its effects on shore after landfall. So just as the Richter Scale was fine for measuring the power of a quake but less useful for measuring its felt damage, leading to the Mercalli Scale for the latter, so demands are now being made for a similar alternative to the Saffir-Simpson Scale.

The arguments behind the demand for a new scale relate to the many factors that affect the impact of a hurricane on shore after landfall. They include wind speed; size of the area covered by the wind; speed of the storm as a whole, something that determines the height of the storm surge; angle of approach to the shore; shape and height of the shoreline; water depth and state of barrier islands, particularly the volumes of sand near shore; and finally, recent storm history, which may have affected all of the above.

Occasionally a hurricane takes an unexpected path because of cold air masses over the interior, which can greatly increase the rainfall. Hurricane Agnes, in

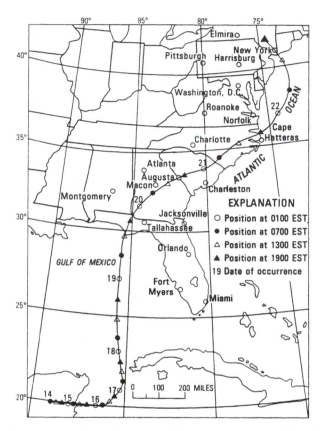

Figure 7.6. The three-day destructive path of Hurricane Agnes in 1972. Hurricanes do not usually travel this way, and as a result, many authorities were unprepared for the heavy rains and flooding that ensued. In addition, forecasting was inaccurate. Much damage was caused by the unexpected westward turn on the twenty-second day.

1972, was one of these (Figure 7.6). It originated in the Caribbean and moved northward across the Florida Panhandle through the Carolinas to New York State. In some places 45 centimeters of rain fell over a two-day period. Many streams experienced peak flows several times greater than the previous maxima on record. In the 12 states affected, 117 lives were lost.

Hurricane Hugo (Figures 7.7 and 7.8) struck the U.S. mainland in mid-September 1989 just north of Charleston, South Carolina, with winds of 215 kilometers per hour and a storm surge reaching six meters. The barrier islands were completely inundated by the storm, and their beaches severely eroded, with sand being either washed landward or carried offshore. Altogether 29 people from South Carolina lost their lives in the course of the storm, and damage to buildings and property amounted to $6 billion.

Where there was a wide high beach and sand dunes, damage was greatly

Figure 7.7. This is one outcome of Hurricane Hugo, the most destructive storm to reach the east coast of the United States prior to 1989. It caused 20 deaths in Charleston, South Carolina. Estimated cost of the damage was $5 billion. These rocks that finally came to rest in the living room were taken from the barrier wall that was designed to protect the home.

reduced, and those buildings that had been built to withstand high winds and flooding survived. By contrast, where, as in Folly Beach south of Charleston, beaches had been severely narrowed through long-term erosion, there was substantial destruction. Homeowners had tried to compensate by dumping boulders and concrete rubble on the beach, hoping they might serve as a retaining wall. They did not.

Every hurricane involves a small army of private and public agencies. The U.S. Geological Service (USGS) provides maps to the Federal Emergency Management Agency (FEMA), the American Red Cross, the U.S. Army Corps of Engineers, and the Defense Mapping Agency. These maps are vital to rescue efforts because they are the only ones that show streams, buildings, topography, and roads. Rescue crews on the scene use these maps to plan their efforts and get supplies to the places of greatest need. In 1992 when Hurricane Andrew struck—one of the costliest disasters on record, with losses estimated at $30 billion—the USGS distributed 5,000 individual topographic maps of Florida and Louisiana and 200,000 maps in bulk shipments.

The USGS and other agencies have played a more and more active role in preparing for future hurricanes. While the probability that any one of these storms will hit land at a given point in a given year is low, and still less likely

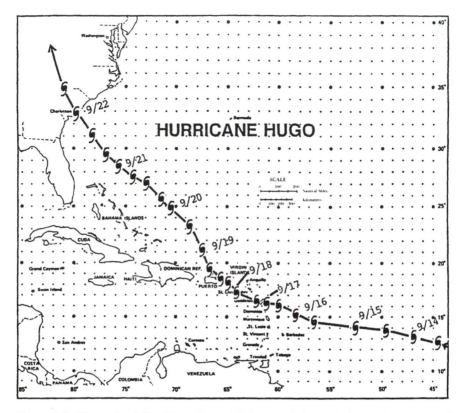

Figure 7.8. The path of Hurricane Hugo, 1989, one of the very destructive storms of recent years. The map shows the path taken, the number of days elapsing from first sighting far out in the Atlantic until it struck Charleston, South Carolina. Most hurricanes travel along routes very similar to this one.

in the case of a category 4 or 5 hurricane, there is the danger that a false sense of security may develop. It can be said, for instance, that for any one building within the general zone of historic hurricanes, a 3, 4, or 5 storm will strike it sometime in its lifetime. It also needs to be said that given the many uncertainties associated with predicting hurricanes, there is no assurance that a second powerful storm will not hit again one year later in the same spot as the previous year.

North Carolina was struck by two hurricanes in 1996. The same rare sequence of two storms happened again in 1999. Dennis struck the coast for several days early in September, and Floyd made landfall later in that same month, causing record floods across eastern parts of the state and damaging shoreline structures. Bonnie, a fifth storm, just as big as the other four, reached North Carolina in 1998. Experts described the first of these five storms in 1996 as a 50-year storm, that is to say, one that would be expected to recur once every 50 years, yet all

Figure 7.9. East Bay Street of Charleston, South Carolina, after an earthquake in 1886. Most of the city was destroyed, and 60 people were killed. The strength of the earthquake was more than 7 on the Modified Mercalli Scale for all of the state, and places as far away as Bermuda and Minneapolis felt it.

five of these storms were of equal strength. Obviously, concepts of 50-year storms need to be reconsidered. Long-term forecasting at the present time must therefore remain as a difficult, if not impossible, task.

EASTERN EARTHQUAKES

Charleston has suffered severely over the years from hurricanes, but this city is also the scene of a major earthquake. In geological language, it is described as quite recent, 1886 to be exact, with the implication that something like it might happen again. The 1886 event was the greatest quake to hit the east coast of the United States in historical times (Figure 7.9). It probably measured 7 on the Modified Mercalli Scale. More than a hundred buildings and many thousands of chimneys in the city were destroyed. About 60 people were killed. The reason for the enormous amount of damage to chimneys and the lesser destruction of buildings stems from an edict dating back more than 40 years following a disastrous city fire.

Older wooden buildings had been burned to the ground in that fire, so for all future construction, brick was required. Unfortunately an inferior type of mortar was used in large numbers of the newer buildings, and these were the buildings that toppled in the earthquake. The quake was felt all the way from Canada to the Gulf of Mexico. Any earthquake in this part of the nation causes damage over a larger area than comparable quakes in the West because of differences in bedrock. Railroad tracks buckled in a number of locations, and telegraph wires were cut, leaving Charleston with no communication links to the rest of the world for a couple of days.

Although there are numerous faults all along the eastern seaboard, the cause of the Charleston earthquake did not lie with any of them; it was pressure from ocean crust moving westward from the Mid-Atlantic Ridge, pushing North America slowly toward Asia and creating tensions in the continental lithosphere in so doing to cause the quake. Pressure from the Juan de Fuca plate as it moves crust in the opposite direction heightens these tensions; it slows the process slightly but does not stop it.

I mentioned that earthquakes could occur again in or near Charleston. Recent research revealed evidence of large quakes near Charleston approximately 600 and 1,300 years ago, suggesting that the frequency of these monster earthquakes may be on the order of every 500 years. After the 1886 quake people reinforced their buildings with wall anchors that tied walls and roofs to the floor in order to prevent them from being blown out in another earthquake. Their effectiveness has yet to be tested, and they may not have to be tested in the lifetimes of the next several generations.

There is widespread indifference to earthquake risk along the eastern seaboard. There are several reasons for this attitude. For one thing, it's the western parts of the continent that get all the attention regarding this particular hazard. For another reason, the oldest records provide information on the large quakes only, and since these are infrequent, the impression gained is that earthquakes are rare events. Since the 1960s this has changed due to the deployment of the Global Seismograph Network, which records small quakes as well as larger ones. Now we know that, on average, a small earthquake strikes somewhere within the regions of the Appalachia and Coastal Plains every day.

The following are some of the larger quakes that have hit the east coast over the past 250 years: Cape Ann, Massachusetts, 1755, magnitude 6; New York, New York, 1884, magnitude 5; Charleston, South Carolina, 1886, magnitude 7.7; Gills County, Virginia, 1897, magnitude 5.8. The most recent prediction about the future comes from the National Center for Earthquake Engineering, Buffalo, New York. In 1990 it reported that there is a very high probability of a magnitude 6 or greater earthquake occurring somewhere in the eastern United States on or before the year 2010.

The GSN is a system of 100 seismograph stations deployed worldwide, collecting data, monitoring earthquakes, and continuing research on the earth as a whole. One unique feature of the GSN is its ability to estimate, within an hour, the geometric orientation and overall length of a fault that caused an earthquake. This is the kind of information needed to assess the quake's damage potential and the likelihood of a tsunami if it happens to be in or near an ocean. Researchers at Columbia University, using GSN data, recently discovered that the earth's solid core rotates at a slightly different rate than the rest of the planet, thus confirming the existence of and changes in the earth's magnetic field.

Several GSN stations are responsible for monitoring the Comprehensive Nuclear Test Ban Treaty between the United States and Russia. In August 1997, a small seismic event occurred near a known Russian nuclear test site on an island in the Barents Sea, and this was monitored by a GSN station in eastern Europe. First reactions were that an underground nuclear explosion had taken place. As additional data came in, it was found that the site of the suspected explosion was not at the test site but in the ocean, some distance from land. The details of the seismic waves, as they were more closely examined, showed that they defined an earthquake, not a nuclear explosion. The GSN has records of past nuclear events and was able to match them with the signals from this particular set of seismic signals.

OTHER HAZARDS

Smog Alerts

Smog is a summertime problem in Georgia, especially in the biggest cities, one that threatens health and disrupts everyday life because car traffic and industrial operations have to be scaled back whenever smog levels are dangerously high. Ozone is the culprit responsible for this health hazard. It is a gas formed in warm weather by a chemical reaction from two groups of chemicals: volatile organic compounds (VOCs) and oxides of nitrogen (NOX). Ozone is a beneficial gas high up in the atmosphere because it shields the earth from harmful ultraviolet rays, but on earth, it's a pollutant, dangerous at certain levels to plants and humans alike.

VOCs come from industries and cars, so government is focused on controlling the amounts of ozone released from these sources. Using public transportation instead of cars helps, but even when there is success in that direction, the gains are offset by growth in population. NOX comes from burning anything, so a range of controls have been initiated: Gasoline-powered vehicles of all kinds are regularly tested for volume of emissions and forced off the road until these emissions are reduced to a safe level; only low-sulfur gasoline is available in the summer throughout a 25-county area in northwest Georgia; and all power companies are required to lower their NOX emissions.

Groundwater Depletion

Water users in the 24 counties of coastal Georgia obtain most of their water from the Upper Floridan Aquifer. Over time withdrawals have reduced pressure in the aquifer, leading to saltwater intrusions. Figures 7.10 and 7.11 illustrate the seriousness of the problem. Between 1958 and 1998 the water in a representative county, measured in distance below ground level, dropped from 15 to

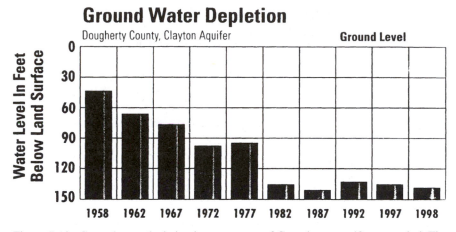

Figure 7.10. Groundwater depletion in one county of Georgia over a 40-year period. The biggest changes came after 1977 with the rapid growth of irrigation agriculture in this area. Subsequently strict regulations were imposed on new permits in order to slow down the rate of depletion.

45 meters. From 1968 to 1998 saltwater increases, measured in chloride concentrations, rose from 200 milligrams per liter to 780.

The biggest drop in freshwater pressure came after 1977 when a rapid growth in irrigation occurred in the agricultural areas of southwest Georgia. Georgia's total water use increased by 20 percent between 1990 to 1995, so as in the case of air pollution through ozone levels, new regulations had to be introduced to limit the encroachment of the sea on the Upper Floridan Aquifer. Beginning in 1997, no new withdrawals were permitted from this aquifer, and new permits for wells were limited. Farther west, no new withdrawals were permitted from the Clayton Aquifer. At the same time, fresh studies were undertaken to identify more fully the nature of the geology and hydrogeology below ground.

Swelling Soils

Soils and rocks that swell or shrink because of changes in moisture are known as *expansive* or *swelling soils* (Figure 7.12). Two rock types are the parent materials: Volcanic rocks containing silicate minerals are one type, and sedimentary rocks containing clay minerals are the other. These soils are found all over the nation in varying concentrations, with the highest in parts of the Great Plains and southern areas of the Coastal Plains. Wherever they are, they do costly damage to roads, buildings, and buried utilities—billions of dollars' worth annually. Most of the 250,000 new homes built on expansive soils every year suffer some damage over their lifetime.

A typical trigger for these soils to expand is the downspout on a home, which can so easily direct water into the soil beneath the foundation of the house. Conversely, a very heavy load on soil can prevent swelling even when large

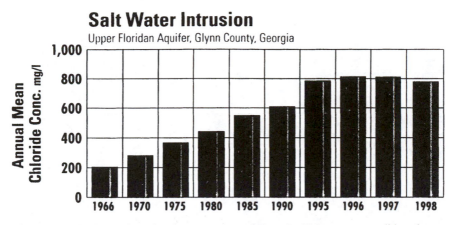

Figure 7.11. Saltwater intrusion in one county of Georgia. This seesaw condition always happens wherever fresh- and saltwater are in close proximity to each other. The degree of salt contamination is measured by the chloride concentration. Because of reductions in groundwater withdrawals, the intrusions level off in the 1990s.

volumes of water are released into the ground. A high-rise building can be built in an area of expansive soils because its weight is great enough to minimize the amount of swelling. What happens is that the restraining pressure from the weight of the building is greater than the swelling pressure developed in the soil. Expansive soils can easily be recognized by lab analysis of the clay-mineral content.

The best way to reduce damage from expansive soils is to avoid them. If that is not possible, then other alternatives can be employed: The offending soil can be removed and fill put in its place, provided the soil layer is not very thick; placing an adequate weight on top of the soil, as in the example I have just described, is another way; a third way is the improvement of drainage around buildings so that little moisture reaches the swelling soils. Blacktopped roads can increase the risk of damage from swelling soils because the capillary action by which water naturally rises to the surface, and is then released into the air, is inhibited by the blacktop.

In central Mississippi and beneath almost all of the metropolitan area of the capital city, there is an outcrop of expansive soil that causes a lot of damage. Yazoo clay here forms a geologic unit 130 meters thick, which expands when wet and contracts when dry (Figure 7.13). Roads and highways, foundations, sidewalks, and structures generally are all subject to damage from this type of soil. Risks are greatest under slopes where soil creep and slides can add a downhill component to any soil movement.

WETLANDS

Once considered almost worthless real estate in their natural state, wetlands have long been treated to draining and filling for farmland and urban expansion.

Figure 7.12. Expansive or swelling soils are found all over the United States, with greatest concentrations in the northern Great Plains and southern Texas. Paved roads like the one shown here, when constructed inadequately on swelling soils, will heave and crack.

More recently people have come to realize that wetlands of all kinds are immensely important to both the environment and the economic health of the nation. They are important as habitats for large and varied populations of wildlife. They reduce the effects of flooding on developed areas, recharge groundwater resources, serve as natural filters for reducing pollutants in water, and support a wide array of recreational activities.

Natural processes of wetlands degradation coupled with the actions of people have reduced the total amount of wetlands in the coterminous United States by 50 percent over the past two or three centuries. These losses are continuing, and nowhere are they greater than in the Mississippi delta plain of Louisiana. Here as much as 100 square kilometers of wetlands are lost every year. Louisiana has one quarter of the vegetated wetlands and 40 percent of tidal wetlands in the coterminous United States, and the losses being experienced come from a combination of natural and human activities.

Conflicts between people and the natural environment have always existed along coasts, and the increasing desirability and accessibility of these places intensify these conflicts. Dramatic population increases in coastal states, much greater than those elsewhere, are apparent in census after census. At the present time, half of the U.S. population live in coastal states. Projections indicate that the numbers in southern and southwestern states will grow by 32 and 51 percent,

Figure 7.13. Uplifted floor in a home in the Jackson metropolitan area, Mississippi, as a result of swelling clays of the Yazoo formation. The floor has taken a generally convex shape, forcing the owners to move flower pots to the center where they can be on a fairly level surface.

respectively, by the year 2025. There are already 34 million people living in Texas and Florida, both likely states for hurricane strikes. As numbers of people in these coastal locations grow, demands increase for transportation facilities, potable water, and arrangements for waste disposal. Pollution is already severe in the larger coastal communities.

The natural hazards in wetlands are severe and increasingly destructive as global warming gives rise to more and more powerful storms. Coastal lands and sediments are constantly in motion. Waves and tidal movements move sand along the coast from place to place. Storms cause erosion in one area and deposit sediment in another. Over time, great changes can take place in shorelines (Figure 7.14). A lighthouse that was built a little over 100 years ago at the eastern end of Long Island and set back 60 meters from the edge of the cliff is now at the cliff's edge and has to be protected from collapse by a seawall. The Cape Hatteras lighthouse had a similar life span and recently was pulled landward to safety from the sea. It had originally been built 500 meters from the sea.

Galveston, Texas, is the site of the worst natural disaster ever to strike the United States. In September of 1900 a hurricane with wind speeds of 160 kilometers an hour created a six-meter storm surge that covered the entire island on which Galveston stands. At least 6,000 people died, and more than 3,500

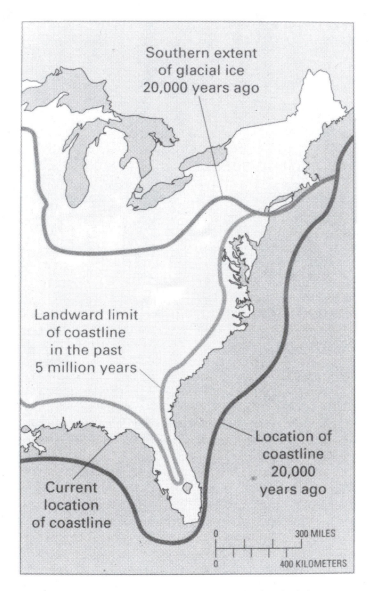

Figure 7.14. Global changes take place in the level of the seas as ocean basins move up or down. During the last great ice age about 36,000 years ago, huge amounts of ocean water were changed into glaciers, resulting in a 100-meter drop in sea level. Sea level is now rising at a rate of about 12 centimeters per century.

houses were destroyed. One third of the city was leveled by wind, waves, and wreckage. Following the disaster, the land surface was raised by three meters and a massive seawall built along the entire Gulf side of the city. Later additions extended the wall along the coast for 16 kilometers.

The U.S. Geological Service and the Coastal Studies Institute of Louisiana State University have been studying coastal erosion and publishing atlases that show their findings, particularly the effects of erosion on barrier islands, which are important protectors of wetlands. These islands date from the different channels in the Mississippi Delta, which deposit the offshore sediments. The Mississippi River has a history of changing course from time to time over the past thousands of years, and each time it moves, the barrier islands become vulnerable to destruction from storms as they are no longer receiving new sediments.

Hurricane Andrew in 1992, one of the costliest on record, stripped sand from 70 percent of all Louisiana's barrier islands. It also smothered 80 percent of the oyster reefs behind the islands with a one-meter blanket of sediment. Earlier in Florida, Andrew completely stripped vegetation from the northernmost Florida Keys and destroyed many old stands of mangrove trees. It needs to be said that hurricanes do not have to reach the barrier islands in order to be destructive. Hurricane Opal devastated Louisiana's barrier islands while it was still almost 500 kilometers away from them.

Hurricane Andrew directly impacted the Mississippi Delta area in Louisiana with sustained winds of 225 kilometers per hour, gusting to 335 kilometers per hour. A storm surge of almost five meters followed the arrival of Andrew, inundating five communities as far as 40 kilometers inland. The eastern wall of the hurricane cut a 50-kilometer-wide swath of destruction to Baton Rouge. In Florida, Andrew damaged the Keys, as has been mentioned. It made landfall near Florida City, 31 kilometers south of Miami, with winds of 225 kilometers per hour and gusts 20 percent higher. As it swept its way into the Gulf of Mexico, Andrew destroyed roads, power lines, and utilities. The cost of all this damage was $20 billion.

Barrier islands and beaches are places of greatest interest. They are the vacation spots, the places where the hotels are and where we find all the knick-knacks. Three out of every four Americans visit beaches when they are on vacation. We build breakwaters, concrete walls, and rock heaps or logs to block waves and keep sand from wandering. We also create artificial reefs offshore to make waves break farther out. All of this may have short-term benefits, but over a longer period of time, they destroy the very thing they were supposed to conserve, the beach sand. A concrete wall, for example, gives energy to a wave, enabling it to carry sand away from the beach into deeper water, to be picked up in due course by waves and carried to another beach.

Barrier islands are vulnerable, as I indicated, because they lose their new supplies of sediment when river channels shift. This is a slow process, not very evident from year to year, so many people find they can build homes on these islands. Folly Beach is a spot on a barrier island off the South Carolina coast

Figure 7.15. This is a levee on the Mississippi River that is beginning to leak. Unless something is done quickly, the leak can quickly become a stream of water and then a scene of catastrophic failure. Sandbags are being placed on an underlying grid to strengthen the levee and stop the leak.

Figure 7.16. Steele Bayou drainage structure in an area south of Mayersville, west of Jackson, Mississippi.

where buildings are raised high enough to be safe from the highest high tides and where people claim they have survived many hurricanes. Erosion by waves and wind is actively at work at this barrier island, lowering its height every year by a fraction of a centimeter. At the same time, groundwater is being withdrawn for domestic use, and gradually this lowers the level of the land as we saw in California. There comes a moment, as happened at Folly Beach in 1989, when the island is so weakened that the next hurricane is the fatal one.

Flooding is a constant concern in a low-lying area like Mississippi. Levees at different heights along the banks of the Mississippi River are silent reminders of this flood threat (Figures 7.15 and 7.16). In the late 1990s a plan was advanced to construct a flood control and pumping station in the southern part of the Mississippi Delta. It would be the largest anywhere in the country, and its main purpose would be to protect agricultural land. People concerned with the protection of wetlands are adamantly opposed to this development, as it would seriously reduce the supply of freshwater to the coastal wetlands and thus allow greater encroachment of saltwater.

Mitigating Damage

This region, the Coastal Plains, demonstrates better than most the need for improved knowledge of the terrain so that human activity can be guided by accurate information. Sinkhole and swelling soil problems can be avoided if ground conditions are known. Fortunately, new efforts are under way to gain a better knowledge base. We saw one example of this in relation to Louisiana's barrier islands. The U.S. Geological Service is involved in many other research projects of this kind. When it comes to landslides and floods, the challenge is to learn from past experience. This is extremely important in anticipating hurricanes because there is every indication that future storms of this kind will be more violent than those we saw in the 1990s.

The typical response to landslides and floods at the present time is to clean up and rebuild, leaving things as they were before the storm. There is a great need to do better than this. Experiences of past storms should be incorporated as part of future planning. Nature should teach us lessons and guide us as we take account of weaknesses in past designs. Too often damaged structures are rebuilt as they were before, only to be destroyed again and again, as happened with Highway 12 in North Carolina. Even if no single house is lost, as happened when Topsail Island was hit, there are many costs involved in the repair of infrastructure, and these should be given consideration before repairs are begun.

REFERENCES FOR FURTHER STUDY

Bascom, Willard. *Waves and Beaches: The Dynamics of the Ocean Surface.* Garden City, NY: Anchor Press, 1980.

Chabreck, R. A. *Coastal Marshes: Ecology and Wildlife Management*. Minneapolis: University of Minnesota Press, 1988.

Fuchs, Sir Vivian. *Forces of Nature*. London: Thames and Hudson, 1977.

Kaufman, Wallace, and O. H. Pilkey, Jr. *The Beaches Are Moving: The Drowning of America's Shoreline*. Durham, NC: Duke University Press, 1983.

King, C.A.M. *Beaches and Coasts*. New York: St. Martin's Press, 1972.

Leatherman, Stephen P. *Island States at Risk: Global Climate Change Development and Population*. Charlottesville, VA: Coastal Education and Research Foundation, 1997.

Morrison, H. R., and C. E. Lee. *America's Atlantic Isles*. Washington, D.C.: National Geographic Society, 1981.

Pilkey, O. H., Jr., and T. D. Clayton. *North Carolina and Its Barrier Islands: Restless Ribbons of Sand*. Durham, NC: Duke University Press, 1998.

United States Department of Commerce. *Hurricane Hugo: 10–22 September, 1989*. Silver Spring, MD: Department of Commerce, 1990.

8

CENTRAL LOWLANDS

The Central Lowlands is often referred to as the heart of America—and with good reason: If we look at the names of the eight states with populations of 10 million or more, this region has three of them, Illinois, Ohio, and Michigan, more than any one of the other five. The other states in the region are Indiana, Iowa, Minnesota, Missouri, and Wisconsin. The distinguishing geological imprint on this part of the nation was the last ice age. There were several ice ages, but the last one created the visible landscapes of today, and while other parts of the country were also affected by massive ice erosion, in no other were all of the states so greatly influenced. Michigan is a good illustration of what happened.

MICHIGAN

Geological formations in this state date back a very long time, some as old as 3.5 billion years. This is not surprising because Michigan, along with several other states in the Central Lowlands, belong to what is called a craton, that is to say, an area of bedrock that is exceptionally stable, changing little over time compared with, say, coastal areas. The rich legacy of ancient rocks provided the state with valuable minerals: Precambrian metamorphosed sediments that were deposited about 2 billion years ago are now the source of Michigan's iron ore mines; volcanic rocks of similar age provide the copper that has been mined here since the 1840s; much later, during the Cambrian period, carbonate rocks along a fault line in south central Michigan led to the formation of a giant oil field.

The advent of glaciers during the Pleistocene epoch was the development that shaped today's landforms. There is evidence of the impact of the last three of the four major glacial advances as these masses of ice moved many cubic kilometers of material to new sites farther south, leaving in their wake as they receded moraines, till, and other deposits. Ice from the last advance moved northward out of the state about 12,000 years ago. During the warmer weather between ice ages, new plants and animals moved northward. Today the average height of the land is approximately 225 meters above sea level.

GEOLOGICAL OVERVIEW

All the other states of the Central Lowlands were as profoundly affected by the last phase of the Pleistocene ice ages as was Michigan. Neighboring states were also impacted to different degrees. There were numerous episodes of glaciation during the Pleistocene time period, and as the ice advanced and retreated, it left a complex surface of glacial lake deposits, moraines, glacial drift, and outwash plains in depths ranging from a few meters to more than 300. Most of the surface deposits that we see today came from the last phase of the Wisconsin glaciation, which began less than 100,000 years ago and finally ended about 10,000 years ago.

The glaciated area was pressed down some distance into the ground, much of it below sea level, due to the weight of the ice, and this land is now gradually rebounding back to its former state. For hydrogeologists, the extensive aquifers within the sand and gravel of the glacial drifts provide a major source of groundwater resources. At deeper levels, in the bedrock, are other extensive aquifers, but these are long-standing features whose characteristics are better known. Water resources in the glacial drifts are relatively new, and the study of them requires a three-dimensional approach, measuring their distribution, thickness, and type of rock.

ICE AGES AND THE GREAT LAKES

The causes of these massive changes we call the ice ages are uncertain, but the most likely one is the changing orbit of the earth around the sun, as I mentioned in my introduction. In any case, immense blankets of ice gradually formed in northern latitudes and moved southward, or rather flowed southward as more and more ice accumulated, pushing the underlying ice forward. It was a process that could be compared with the action we have seen at midocean ridges as more and more magma is pushed up and out from inside the earth. By the time the ice sheets reached their maxima, almost all of Canada and much of the northern United States was covered with ice, in many places as much as 2 kilometers thick.

Some scientists see our present situation as a pause between ice ages. It is quite clear to them that the various periods of ice were followed by warmer climates lasting as long as 20,000 years, with temperatures like those of today, before another phase of ice swept over the land. There are ice sheets on Greenland and Antarctica today, yet these places once had very different regimes, including tropical climates at times. In time we may be able to predict when the next ice age will arrive. For now there is no sure way of predicting it. We just know that we will have plenty of warning of its arrival.

Where did all the moisture come from to create an ocean of ice? From the sea. That was the only possible source since the worldwide hydrologic cycle is a closed system. Water may change its form—solid, liquid, or vapor—but the total quantity remains the same. Taking water from the ocean in such enormous volumes lowered sea level by about 100 meters all around the east, west, and gulf coasts, even though some downward pressure of the ice on the land partly compensated for the lowering of the sea level. What is now continental shelf was then land for thousands of years, and today the evidence for that is found in old tree stumps underwater on the continental shelf. Following the retreat of the ice, sea levels rose, and land also began to recover its former state.

How do we know what the ice sheets did and when they came? One set of clues is found in the bedrock over which the ice moved. Rock fragments at the base of the ice sheets scratched and grooved the underlying rocks, and these marks, still readily seen today, told the direction of ice movement and, by tracing the marks back to their sources, the places from which the ice came. The oldest deposits left by the ice when it started to retreat can be examined by radiocarbon dating methods, and thus an approximate time is obtained for that period.

The scale of the erosion and deposition that took place needs further explanation. The amounts of material that were transported are quite astounding, millions of cubic kilometers of soil and rock. Huge blocks of rock were sometimes plucked from bedrock, perhaps where a local fault caused some weakness, and carried along embedded in the base of the ice mass, to be deposited at the melt point as isolated boulders, different in appearance from the surrounding rocks. These boulders are known as erratics. Where the path of the ice coincided with the direction of flow of a river valley, the shape of the valley changed from a "V" to a "U."

Glacial lakes were formed wherever deposits blocked the escape routes of meltwaters. The largest of these, Lake Iroquois, was one of the early forms of what is today the Great Lakes. It occupied an area larger than present-day Lake Ontario, and its southern border, a beach ridge of sand and gravel, can still be seen close to the southern border of Lake Ontario. Meltwaters from these glacial lakes completely modified historic patterns of drainage, creating new waterways and new lakes. It would not be too much to say that everything that distinguishes a landscape—its vegetation, drainage, and surface features—which had evolved over many thousands of years, was completely changed by the ice.

The area we now know as the Great Lakes was a lowland territory drained by several major streams. It is a big area of water today, 240,000 square kilometers for the surface areas of the lakes alone, holding 10 percent of the world's total amount of freshwater. It is greater than any other single freshwater lake in the world. All around the lakes are extensive stretches of first-class agricultural

land, and it was these good farming areas that attracted the first settlers in earlier
times, laying the foundation for the present huge populations of people and
industries.

As ice sheets advanced and retreated, the weaker rocks fell victim repeatedly
to the scouring action of ice. The eroded rock and soil were carried farther south
and deposited at the southern edges of the ice as terminal moraines. These
moraines, together with the excavated areas to the north, defined the basins of
the lakes that formed as the ice retreated and meltwaters took their place. For
a time, overflow water from these lakes was carried southward by rivers into
the Mississippi Basin.

The gigantic excavations that took place as glaciation after glaciation arrived
repeatedly lowered the level of the land so that today the bottoms of all but one
of the lakes are below sea level. The bedrock of shales and limestones was
eroded and the rock fragments carried southward to be deposited in moraines
and other landforms. We will look at Indiana to see the results of the ice ages
in that state, and one of the evidences will be the presence of unconsolidated
quantities of these shales and carbonates.

The process of ice advance and retreat, followed by different patterns of lakes
and moraines at the end of each phase, meant that there were times when large
tracts of land were under water for thousands of years. These lands today are
flat areas of excellent soil, ideal farming country. The southern limits of the ice
advances were determined by temperature: Termination occurred when the rate
of melting equaled the rate of arrival of new ice. At that location, temperatures
were very close to the freezing point, like temperatures today in southern Alaska
at the melting point of glaciers.

INDIANA

Indiana provides a good illustration of the impact of the Pleistocene ice ages
on the states south and west of the Great Lakes. About half of the state, the
northern half, was largely shaped by the final phase of Wisconsin glaciation,
whereas the southern half was much less affected. Few bedrock exposures are
seen in the north because they are covered over with thick layers of glacial
debris. The south has some layers of glacial material, but they are thin and in
some places completely absent. Most of the southern counties have bedrock
outcrops, the youngest of which are the Pennsylvania strata, the location of
Indiana's coal resources (Figures 8.1 and 8.2). There are practical consequences
arising from these contrasts: Agriculture is dominant in the economy of the
north, whereas mining is more important in the southern half of the state.

Throughout the state, though more evident in the northern areas close to the
Great Lakes, the legacy of the ice ages takes the form of glacial till, that is to
say, a mixture of debris ranging in size from boulders to fine silt. Sand and
gravel occupy some places where floodways carried water away from the ice
sheets and where streams and rivers now flow. Elsewhere the outwash deposits

Figure 8.1. Friar Tuck coal mine, Indiana. A deposit of waste material from a coal-preparation site that has been eroded by wind and rain. Photo taken in the 1980s.

form vast plains. Farther north near the southern shore of Lake Michigan sand dunes dot the landscape, the result of erosion along the western shores of the lake, a subject that will be discussed later on in this chapter.

The various glacial sediments have had great economic significance in Indiana both in earlier times and today. Active and abandoned gravel pits are everywhere. Clay from glacial lake sediments as well as glacial till is used in the manufacture of brick, tile, and ceramic products generally. Muck and peat from ancient glacial lakes serve as soil substitutes. Quartz sand deposits from the vicinity of Lake Michigan are used in the manufacture of glass. Crops are grown in the soils derived from glacial debris, and highways and buildings use various kinds of soil and rock materials.

SURFACE EROSION

Lake Shores

Wave and tidal action are not normally serious agents of erosion on lakes (Figure 8.3) because they are relatively small bodies of water. Lake Erie's size and oblong shape make it an exception to this generalization. Communities like Buffalo at one end and Toledo at the other suffer from strong storm surges that can quickly raise the level of the water by a meter. Although these surges are short-lived, they do considerable damage, primarily from flooding.

Figure 8.2. Subsidence in a farm field above an abandoned underground coal mine, near Washington, Indiana. Photo taken in the 1980s.

Presque Isle is a popular state park near Erie, Pennsylvania. Millions of visitors go there every year. It is a 10-kilometer-long sandspit that built out from the Lake Erie shoreline, and it remains connected to that shore by a narrow neck. This neck is disappearing at the rate of 2.5 meters per year, probably due to a classic case of human interference without adequate awareness of the consequences. Jetties were built on the Ohio shore of Lake Erie, and it is felt that they interfered with the eastward-moving longshore currents that would normally replenish the Presque Isle neck with fresh sediment. The U.S. Army Corps of Engineers is now working on a costly plan that uses segmented breakwaters in order to stop the erosion of the island's link with shore.

Beaches and cliffs are frozen in winter, so there are no storm waves, but in both spring and winter, ice formation and thawing can accelerate flooding. Evaporation rates are reduced, and this condition plus other factors such as snowmelt can make a difference of 30 centimeters between winter and summer water levels. Additionally, there are the long-term cycles of high and low levels of water in the lakes. A period of high levels in the 1980s threatened the protective engineering structures along Lake Michigan's waterfront at Chicago, and this led to a major study in 1987. Geologic structures, lake-level studies including prehistoric fluctuations, and examinations of bluff recession and evolution of beaches all were included in this study.

Wide fluctuations in the levels of Lake Michigan, as much as three meters, have been traced back to both prehistoric and presettlement times. In the 1600s

Figure 8.3. Bluff erosion on Lake Ontario, 1996. These bluffs are composed of glacial deposits, and they are continuously failing when strong wave action removes these deposits during storms. The saturated zone at the base of the bluffs is the place where erosion is most active.

there is a record of one of these large shifts in water level before the area was extensively settled by Europeans. By comparison, the difference between the 1964 low level of the lake and the 1986 high is only half that of the earlier ones, so the implication is that higher fluctuations as a result of climatic factors will likely occur again in the future. The impact of global warming on these changes is still uncertain.

Findings on bluff retreat revealed a range of erosion rates from 10 centimeters near Chicago to 3 meters in areas 60 kilometers farther north on the west side of the lake. Sediment eroding from the bluffs provides a sand cushion along the shore that partly protects the shore from further erosion. However, as people continue to build structures to protect the bluffs from getting worn away, the nearshore sand cushion gets smaller and smaller. As a result the underlying finer layers of the lakebed are exposed to wave attack, and coastal retreat, instead of slowing down, accelerates, one more example of human interference without adequate knowledge of the consequences (Figures 8.4 and 8.5).

Ice ridges that build up along the lakeshore were formerly thought to be protectors of the shoreline. Now we know that this is not the case. Substantial quantities of sand are trapped in the ice and carried away either alongshore or out into deep water at breakup, thus increasing erosion. Much of the sand finally ends up in the dunes of coastal Indiana at the Indiana Dunes National Lakeshore.

Figure 8.4. Niagara Gorge in the early 1970s. A power plant was under construction when this block fall occurred, destroying the plant. Niagara Falls is an important source of electrical power.

Sand accumulating on broad beaches is blown on shore by strong winds from the north to form these dunes. Most of the sand seems to have moved southward along the eastern shore of Lake Michigan, rather than the western.

At the eastern end of the Great Lakes, the southern shores of Lake Ontario are composed of drumlins, glacial deposits from the last ice age. They are therefore recent deposits, not very well consolidated, and so are subject to rapid erosion. This is soon obvious to an observer: The bluffs at the water edge are high, as much as 50 meters above lake level, with slopes of more than 45 degrees. Deep ravines can be seen in the sides of the bluffs, the joint result of erosion by weather and undercutting by waves. The end result, unless some remedial measures are taken, will be bluff recession and loss of land, just like the western shores of Lake Michigan.

FLOODING

Central Lowlands Flooding

Floods are the most destructive hazard facing the people of the United States and the most ruinous to life and property. Nowhere is this hazard greater and more widespread than in the Central Lowlands. It is here that the biggest concentration of rivers and the highest volumes of water are found. The Missouri, Mississippi, and Ohio Rivers all coalesce to form the mighty Mississippi (Figure

Figure 8.5. Lake George, New York State, shoreline erosion. Cliff or bluff erosion on a relatively small lake is not common because wind speeds are unable to rise high enough to cause damage. However, if the surficial rocks are weak or unconsolidated, destruction of the kind seen here will take place.

8.6). From Pennsylvania to Montana and from Minnesota to Louisiana, this basin drains 40 percent of the coterminous United States. I am therefore giving more space to the problem of flooding in this chapter than in any other. Floods strike us in myriad forms, but by far the most frequent are the inland, freshwater floods that are triggered by rain or melting snow or through bursting of structures like dams that were built either to protect places from damage or to utilize the water in some way.

An inland, freshwater flood is defined as any abnormally high level of water in a river, forcing it to overflow its banks and cause widespread disruption and damage to crops and buildings. Flooding is a natural characteristic of rivers; floodplains are natural reservoirs and temporary channels for floodwaters. Some years are exceptional as regards the levels of water in major rivers and their widespread distribution across the country. The year 1993 was one of those, but it must be added that the nation is overtaken by less destructive but serious floods at many other times. Early spring floods (Figure 8.7) are common in the northwestern states, the Great Lakes area, the Missouri River Basin, and the Red River of the North Basin.

We saw in the Coastal Plains that a 100-year flood is not always what it is supposed to be, an event that occurs only once in a hundred years. That is the statistical meaning of the term, namely, that on the average such a flood will repeat every 100 years. Hydrologists nevertheless continue to use this as well

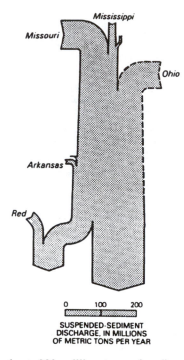

Figure 8.6. Every year about 200 million tons of sediment are discharged by the Mississippi River into the Gulf of Mexico. While this amount is less than it was a few decades ago, it still represents an enormous loss of soil from the river basin. At the same time, it is a vital resource for the wetlands at or near the delta.

as a 10-year flood term because, despite examples like Coastal Plains, these names still define probability for the vast majority of rivers. The USGS maintains a network of 8,000 continuous-record gauging stations in order to be able to warn people and places of an approaching 10- or 100-year flood.

Flood data is only one item of information collected at these gauging stations. Rates of water flow at different times are vital for communities downstream and are widely distributed to public and private agencies. They use the data for design of bridges, channel capacities, roadbed elevations, floodplain zoning, and flood-warning systems. When it comes to flood warnings, the National Weather Service uses all of this information to forecast the stage and time of an imminent flood. In most years there are significant floods in a few locations. Once in a while, there is one of such magnitude and widespread distribution that it is classified among the greatest floods of the nation. The events of 1982–1983 constituted one of these.

Winter rainfall and temperature patterns began prematurely in the late fall of 1982. Early in December and again toward the end of the same month, two storms that were caused by the presence of a very low pressure system over the

Figure 8.7. Spring flooding of the Missouri River downstream from the mouth of the Osage River Junction in 1973. That was the year when a major flood hit the whole of the Mississippi Basin, resulting in the loss of 33 lives and causing damage amounting to more than a billion dollars.

Southwest generated a series of storms and brought heavy rainfall to central and northeast sections of the interior. The slow movement of these storms northward produced tornadoes, severe thunderstorms, and long-lasting heavy rainfall. High water levels in the Mississippi and its tributaries ensued, sustained through December by occasional rains. In all, an area of 400 by 1,600 kilometers was affected, extending all the way from the Great Lakes to the Gulf of Mexico.

Severe damage was experienced in Illinois, Missouri, and Arkansas from the early December storms. Those that came later in the month did major damage to Louisiana, Mississippi, western Tennessee, and Kentucky. Other states farther to the east, while they were not flooded, had very high rainfall. Many of the outstanding peak flows were on the larger rivers because of the widespread and long duration of the rainfall. The mainstream of the Mississippi had high peaks from Illinois down to its mouth. Mississippi State was on the eastern fringe of the storms, so it escaped most of the flooding. Only the western areas were affected.

In several states, streams exceeded previously known maximum floods. Illinois and Missouri each saw this happening in several streams. The southern part of Arkansas, which had been hit by both the early and late December rains, similarly saw all previous records broken. In one location a previous high, dating from 1936, was exceeded by two meters. It was almost the same story in one stream of Louisiana where conditions generally were rated above the 100-year level. Once again, we see examples of 100-year events occurring much sooner than expected. Perhaps the influences of El Niño and La Niña, described below, explain these anomalies.

In April and May of 1983 heavy flooding was again experienced in the middle and upper Mississippi, partly the result of further storms of the kind that drenched the area in the previous December, partly due to snowmelt. Once again there were record-breaking floods. The Pearl River in Mississippi exceeded its previous highest, set in 1874. It experienced extensive flooding for the third time in five years, and one of these was calculated to be far in excess of the 100-year level. In total, in the lower Mississippi, for the six months from December 1982 to May 1983, 150 centimeters of rain fell, three times the normal amounts for this period.

Flood of 1993

Flooding in 1982 and 1983 was a serious event, classified as one of the greatest in the nation's history. The devastating 100-year flood in Georgia as a result of Tropical Storm Alberto in 1994 was another example of severe flooding, and no one in that state wants to see a storm of that kind for a hundred years or more. The flood of 1993 was also a 100-year flood but with a difference: Nine states were affected, and the damage inflicted on property and agriculture was $10 billion, the greatest in the history of the United States. In addition, tens of thousands of people were forced from their homes. People still look back on June, July, and August of that year as the "Summer of the Flood."

There were a series of exceptional climatic factors behind the events of that summer. A large low pressure area in the Southeast brought warm, moist air into the upper Midwest, where it collided with colder, dry air, triggering repeated severe rainfalls. All or parts of nine states were inundated with rain, flooding many of their streams. All-time record flows were recorded in seven states— Minnesota, Nebraska, Iowa, South Dakota, Kansas, Missouri, and Wisconsin. The timing of the summer storms was a critical factor. Floodwaters from storm runoff in northern parts of the Mississippi Basin coincided with later storm runoff from tributaries lower down in the basin. The result: huge peak discharges on the lower Mississippi.

The recurring storms throughout the Mississippi Basin resulted in long periods when rivers were above flood stage. At St. Louis, Missouri, the river was above flood level for more than 10 weeks. Runoff at this point was consequently many times above normal for all of July and August. Midway in this period water level was two meters above the previous highest. The hydrologic and social impacts of all this were severe and widespread. While most of the federal levees stood firm, a few failed. One of these failed levees left 250,000 people in Des Moines, Iowa, without drinking water for 19 days. Other cities experienced similar disruptions. Thousands of acres of land were inundated. Damaged highways and submerged roads interfered with transportation in the whole flooded region.

Some bridges failed because the fast-moving water scoured out their channel beds and undermined the bridge supports, just as happened in Connecticut. At State Route 51 in Chester, Illinois, more than six meters of scour was found around one bridge support. The Mississippi and Missouri Rivers were closed to navigation before and after the flooding. Millions of acres of productive agricultural land remained under water for weeks during the growing season, and sediment was deposited over the land. Industrial areas, too, were inundated, and there was concern about chemicals being released into the groundwater.

Levee breaks in agricultural areas concentrated the river energy at these sites, causing further damage to levees, deep scour, and extensive sand deposition on the floodplain. As much as 5 percent of the floodplain around St. Louis suffered major damage by these processes. The power of water is proportional to its speed. If the speed is doubled, the erosive power is increased by much more than twice what it was before. Thus when breaches of 500 or 1,000 meters are made in a levee at a time of peak flow, as happened around St. Louis, the enormous volume and speed of water cut channels in the agricultural soil next to the levee, often as deep as 1.5 meters and extending downstream for two kilometers.

These deposits from levee breaks are much thicker and more widespread than the ones that come from overflow of riverbanks. The 1993 deposits from these breaks covered more than 30 percent of the floodplain in contrast to the 5 percent coverage from average overflows. Levee deposits at times are so thick that they mask the features of the preexisting floodplain surface. In places there were deposits in the form of three-meter-deep crescent-shaped sand heaps. Postflood analysis recognized that these dramatic events of nature cannot be fully controlled, and so the risk of flooding is always there.

The costs of these events are now so high that it is almost impossible to share them among those concerned without an aggressive program of damage avoidance or mitigation. A new conceptual basis for floodplain management is needed, one that addresses several questions: where, how often, and how will particular places be flooded? How energetically and with what concentration of force? What is the likelihood of structural damage, extensive scour, and deposition? All this will entail detailed geologic, geomorphic, and hydrogeologic assessments of floodplains in order to give guidance to all the agencies concerned.

El Niño, La Niña, and Floods

These two Spanish terms are names for "The Christ Child" and "The Girl," respectively, but they mean quite a lot to us because they are linked to major weather changes and to floods in the Mississippi and its tributaries. At intervals of 5 to 10 years, there is a disturbance in the ocean and atmosphere in the eastern Pacific Ocean, and the effects are felt all over the world. The ocean waters there get warmer than usual, and because the change occurs around Christmastime, it was given the name El Niño. The warm waters spread northward as far as the Oregon coast, with one effect being a southward movement of the jet stream, the high-altitude air flow that guides cyclonic storms across the nation.

Frequent and intense invasions of cyclonic storms ensued, bringing heavy rain to coastal areas and heavy snow on western mountains. The rainfall in these areas was twice the normal in January and February of 1983, a condition we have already noted. In California, record-shattering rains and coastal storms came ashore and swept across the country. Again in 1992 and 1993, there was another episode of unusually warm water off the coast of Peru, and—as we have seen—there was extraordinary flooding all across central parts of the United States.

By the spring of 1994, the fury of these storms subsided, and there was a period of normality for several years. The name La Niña is given to this stretch of time. It marks a strengthening of the subtropical high and, with it, stronger trade winds. The warmer waters are thus pushed westward, drawing up cooler water from below all along the South American shores. All is well until the next cycle of El Niño occurs.

EARTHQUAKES

Earthquake risks were described briefly in relation to New York State. The general impression with people in that part of the country is that eastern areas of the United States have little to fear from earthquakes. The big ones, they say, are all in the West. We saw how unrealistic that viewpoint is when we studied the Coastal Plains and examined the events around Charleston in 1886. The earthquake that shook that city was just as powerful as any one of those that have hit California over the past 50 years. There may be gaps of several hundreds of years between quakes in that part of the nation, but it is certain that there will be others in the future, just as powerful as the 1886 one, and there is no way of predicting any one of them.

Increasingly, it is becoming clear that land areas in the middle of large tectonic plates can be struck by earthquakes just as destructive as those that occur at the junctions of plates. The fact that they are usually less frequent and that they are not accompanied by volcanic eruptions should not allow us to be less concerned. Recent research findings from the site of the New Madrid earthquakes of 1811 and 1812 reveal a zone of seismic activity unmatched anywhere else east of the Rocky Mountains. Here were experienced, in those years at the beginning of the nineteenth century, the largest historical earthquakes to have occurred anywhere in the coterminous United States. Their strengths were estimated to be at the top end of the Richter Scale at times and between 10 and 11 on the Modified Mercalli Scale.

The first of the 1811, 1812 series of quakes occurred at night on 15 or 16 December 1811. Buildings collapsed, trees toppled, and the Mississippi River changed course. Over the following three months, the area was rocked by two more quakes as powerful as the first one and by many smaller ones. The shock waves rang church bells in Washington, D.C., and both neighboring states such as Indiana as well as distant points such as Boston felt the impact. One seismologist from the University of Memphis predicts that there is a 90 percent chance of a magnitude 6 earthquake occurring in the New Madrid zone within the next 50 years. This is a reasonable prediction given that quakes of this strength have already occurred in 1843 and 1895 and that since 1974 some 3,000 earthquakes have hit the area.

These earthquakes were centered on an area of the Mississippi embayment close to where several states meet—Missouri, Arkansas, Kentucky, and Tennessee. There is a fault zone here associated with the ancient Reelfoot intraplate rift that runs alongside the river. Any repeat of the powerful quakes that struck New Madrid would be a terrifying experience. Today there are more than 14 million people living within the area that was devastated by the 1812 earthquake. Some comparisons can be made with the San Francisco quake of 1906, about which we know much more. Total population affected there was 4 million. Furthermore, because of the nature of the earth's crust in the central parts of the United States, the geographic area affected by any earthquake is bigger.

Precise locations of both the intraplate rift and epicenters of the four big New Madrid earthquakes of 1811 and 1812 have now been delineated. The rift is 200 kilometers long, and the epicenters show a northeast-southwest trend. Another series of events like these would be felt from Denver to New York City and damage buildings in eight states. All of the western two thirds of Tennessee was hit by the nineteenth-century quakes. The state already has its share of other hazards such as landslides, subsidence, and flooding as will be evident from Figures 8.8–8.10. Mass movements are most frequent along the bluffs adjacent to the Mississippi River Valley, where slump and block glide occur in the loess masses. Tennessee and the other three states bordering this earthquake zone would be devastated by any repeat of the 1811 and 1812 events.

Development of the Reelfoot rift began about 600 million years ago. The convection currents that cause the massive tectonic plates to move will also cause them to rip apart. I explained this in Chapter 1 in relation to the spherical shape of the earth's surface and therefore the differential rates of motion of the ocean crust as it moves away from the spreading ridges. The North American continent is impacted on its east coast at different rates, and so internal stresses are created. The latest phase of stress in the New Madrid zone from about 70 million years ago is one of compression, resulting in numerous earthquakes as the squeezed bodies of rock grind past each other along the deeply buried faults. It's a movement similar in some ways to what happens at the San Andreas Fault.

Studies of the New Madrid zone by geophysical surveying techniques show

Figure 8.8. Landslide in Lawrence County, Tennessee. Limestone and dolomite are both prevalent in this county and its surrounding area, so solution by groundwater is always undermining the cohesion of the soil.

Figure 8.9. Flood scene in Davidson County, Tennessee. Few areas of the state are free from the danger of floods. Whenever there is heavy rain, there is a rash of flooded areas. Here the highway had to be closed down.

that earthquakes like the 1811, 1812 ones occur on the average every 400 or 500 years, a figure that is close to the estimates for a repeat of the Charleston quake. It seems that both areas are affected by the same types of stresses, those caused by the differential actions of the crust from the Atlantic's spreading ridges. There is convincing evidence that a very large quake occurred in the New Madrid zone sometime between 1400 and 1670. Data from old Native American sites proved very helpful in identifying the ages of rocks at various depths. For example, potsherds and other organic materials from two ancient cultures, approximate ages 500 B.C. to A.D. 800 and A.D. 800 to 1670, were unearthed and dated by radiocarbon methods.

WATER QUALITY

Thus far we have looked at problems associated with large quantities of water, the destructive effects of flooding. There is another and very different problem about water, the contaminants it carries. We need to know that the water we drink and also use in hundreds of other ways is safe and not a threat to our health. Farmers, too, need to know that the groundwater and surface waters they use will not damage crops. In Iowa sustainable agriculture and improved water quality are top environmental issues. They are closely tied together because

Figure 8.10. Sinkhole in Montgomery County, Tennessee. The white coloring of the underground strata is a reminder that limestone, dolomite, and other carbonates are found in many parts of the state, and these are the rocks that cause sinkholes.

fertilizers and pesticides ensure high crop yields, but they also lead to contamination as runoff carries them into the watercourses. Iowa battles with the challenge of reducing the use of pesticides and fertilizers while still maintaining high crop yields.

The Mississippi River and its tributaries drain 40 percent of the coterminous United States, including most of the agricultural and industrial areas. By the time it reaches the delta region around New Orleans, it is carrying a range of pollutants in the 400 billion gallons of water it pours into the Gulf every day. Many of these are well known as to sources and quantities, but by the late 1980s, the growing concern over what we do not know about these contaminants led to a special study of the river by the USGS. The Ohio Basin contributes half of the total water, compared with one eighth from the Missouri Basin, even though Ohio drains an area less than half that of the Missouri.

Findings from the USGS study were published in 1995. They provide useful information on both sources and levels of pollutants. Bacteria from human and animal wastes were generally found to be too high for recreational activities, but fortunately these contaminants survive only briefly in the river. Wherever high counts of biodegradable detergent were found in the same locations as equally high counts of coliform bacteria, the combination of the two proved conclusively that the cities or communities nearby had failed to give complete treatment to their wastewater.

Caffeine is an interesting element because it distinguishes human waste from all others because it comes from soft drinks or coffee. The amount of this contaminant indicates the degree to which domestic sewage is being diluted by the river. Caffeine content in the Mississippi turned out to be extremely low, less than one hundredth of the amounts found in municipal wastewaters, a helpful reminder that pollutants from domestic sources are not the biggest culprits. Other toxic chemicals in the river could also be traced back to their sources, and so authorities could focus their efforts to reduce the amounts of these harmful elements. One organic chemical contaminant, representative of industrial contamination and found in large quantities in the Mississippi, fortunately is only one quarter as concentrated here as it is in several European rivers.

Two pollutants, PCBs (polychlorinated biphenyls) and lead, received careful study partly because of their toxic content and partly because of the way they mix with the sediments being carried along in the water. PCBs and lead adhere to sediment particles much more than other things, and so they are carried farther downstream. The quantities of sediment found at particular points in the course of the river are thus important indicators of the presence of PCBs and lead. Lead content is now decreasing generally because of the use of lead-free gasoline. At its mouth the Mississippi discharges about 210 million tons of sediment every year.

Agricultural chemicals enter the Mississippi from numerous points along its course, usually in spring or summer when crops are at important stages of growth. Every farmer is concerned about the problems of fertilizers and pesticides. He knows they are necessary, but he is also anxious to have water of high quality, so there is a certain conflict between the two goals. The problem is particularly acute in monoculture, where fields are constantly used for a single crop like corn. In such circumstances, soil quality is degraded over time, and more and more fertilizers have to be introduced to maintain productivity.

Nitrate in the river comes from fertilizers. Its concentration depends on the season of the year and on prevailing rainfall and runoff conditions. The highest concentrations are found in the upper reaches where the biggest farming lands are located. Below the confluence of the Ohio and Mississippi Rivers, these concentrations drop off. It is a similar story with atrazine, a widely used herbicide on farms. The highest quantities are found near St. Louis because of inputs from the Missouri and Illinois Rivers. There are times in the course of the year when this particular contaminant exceeds the maximum allowable level of three micrograms per liter.

Industrial pollutants are the other threat, along with agriculture, to water quality generally. Two important chemicals, used in flame retardants in the manufacture of polyurethane foams and textiles, come into the Mississippi system almost exclusively from the Illinois River Basin. Since its source is so well defined and because it has a long life in solution, it serves as a valuable indicator of all waters from the Illinois River. Another chemical that comes almost exclusively from the Kanawha River Basin in West Virginia also serves as an indicator of waters that reach the Mississippi from the Kanawha via the Ohio.

Mitigating Damage

The shores of the Great Lakes are vulnerable to erosion because they are bordered by young, unconsolidated glacial deposits from the last ice age. Some of the measures being recommended to slow down the rate of erosion include the following: provide alternative outlets for surface water behind the bluffs; reduce the steepness of the cliff slopes to prevent mass wasting when wave undercutting causes cliff collapse; and stabilize the sediment of the bluffs by planting native vegetation. All three of these are costly measures, and the value of their implementation has to be assessed against the costs of the losses that will certainly come if no action is taken.

REFERENCES FOR FURTHER STUDY

Allen, T.F.H., and T. W. Hoekstra. *Toward a Unified Ecology.* New York: Columbia University Press, 1992.

Burton, I. *Types of Agricultural Occupance of Flood Plains in the United States.* Chicago: University of Chicago, 1962.

Cooke, R. U., and J. C Doornkamp. *Geomorphology in Environmental Management.* Oxford: Clarendon Press, 1990.

Kates, R. W. *Hazard and Choice Perception in Flood Plain Management.* Chicago: University of Chicago, 1962.

Marriott, S. B., and J. Alexander, eds. *Floodplains: Interdisciplinary Approaches.* Bath, IL: Geological Society, 1999.

Mickelson, David M., and J. A. Attig, eds. *Glacial Processes, Past and Present.* Boulder, CO: Geological Society of America, 1999.

Morse, D. F., and D. Brose. *Archaeology of the Central Mississippi Valley.* New York: Academic Press, 1983.

Parry, M. L. *Climate Change and World Agriculture.* London: Earthscan Publications, 1990.

Penick, J. L. *The New Madrid Earthquakes.* Rev. ed. Columbia: University of Missouri Press, 1981.

Scott, R. F. *Principles of Soil Mechanics.* Palo Alto, CA: Addison-Wesley, 1963.

9

GREAT PLAINS

The Great Plains extends from the Dakotas to Texas and includes parts of 11 states: Arkansas, Colorado, Kansas, Montana, Nebraska, New Mexico, North Dakota, Oklahoma, South Dakota, Texas, and Wyoming. Topography is flat to gently rolling terrain, the remnant of a vast plain that originally stretched from the mountains eastward beyond the Missouri River. Regional uplift caused streams to cut downward, eroding the plains and isolating them from the mountains. Only in Wyoming does the plain still extend to the mountains.

Windblown sand and silt, derived from the bed of rivers that eroded the plains, were deposited over large areas. The largest expanse of sand deposits and dune topography to be found anywhere in the Western Hemisphere is in west-central Nebraska. The water table is close to the surface in these sands. Most of the Great Plains have a dry continental climate, one of abundant sunshine, frequent winds, moderate precipitation, and high rates of evaporation. Persistent winds and high summer temperatures cause such high rates of evaporation that little is left over to recharge the groundwater system. Only in sand dune areas where water can percolate down to the water table can some precipitation be retained.

GEOLOGICAL OVERVIEW

Groundwater is stored in near-surface deposits of the Tertiary and Quaternary ages, whereas beneath are the bedrock units that range in age from Permian to Tertiary. These bedrock units are, for the most part, impermeable. The oldest rocks that are in contact with groundwater in the High Plains Aquifer are of Permian age, and they occupy about 12 percent of the aquifer under Kansas, Oklahoma, and Texas. These rocks consist of red beds and evaporites that were deposited in extensive, shallow, brackish to saline seas that were created by

Figure 9.1. Morrilton, Arkansas, about 75 kilometers northwest of Little Rock, clearing away a landslide that had blocked Interstate 40.

periodic influxes of marine water. Permian rocks are moderately permeable, so much of their saline content can enter the aquifer.

The Pleistocene ice ages reached south only as far as northern Kansas, but their influence was felt throughout the whole of the Great Plains. (Figures 9.1 and 9.2). Oklahoma's surface was sculptured by major rivers from the glacial meltwaters and by the increased precipitation associated with glaciation. It was a similar story in other parts of the region. The major drainage systems of today were formed during the last phase of the ice ages. In their early stages the rivers often changed courses over the long periods of glacial action, and alluvial deposits from each new location are visible today in alluvial terraces. These are often as high as 100 meters above the floodplains of the present rivers, an indication of the ice sheets' massive scale of erosion.

ARKANSAS

This state, though bordering two other regions, Central Interior and Coastal Plains, is a major player in the water resources and water quality of the Great Plains. The state may be divided topographically into highlands and lowlands. Based on elevation and geology, the boundary between these major geographic regions extends diagonally through the center of the state from northeast to southwest. In the north and west the highlands consist of the Ozark Highlands and Ouachita Mountains, and both of these highlands continue into Missouri

Figure 9.2. Lake sediments from a former northwest glacial lake, now bordering a river.

and Oklahoma. To the south and east the lowlands form part of the Coastal Plains. I have selected this state because of the unique role of the Arkansas River as an aquifer-stream system, a system that is vital to an understanding of the aquifer-based sources of water throughout the Great Plains.

The Ozarks in northern Arkansas are flat topped with narrow ridges and steep-sided valleys. One area, the Salem Plateau, is an undulating plain with average elevations of 250 meters. Other plateaus rise to heights of more than 600 meters. Bedrock is dolomite, sandstone, or limestone. The Ouachita Mountains feature east-west trending ridges and valleys that were formed by the erosion of folded sedimentary rock. Elevations are higher here, up to 850 meters in one area. The Coastal Plain of Arkansas consists of flat bottomlands and low rolling hills. Soils are predominantly sand and gravel. Some of the alluvial plain of the Mississippi occupies the eastern third of the state.

Arkansas River Aquifer-Stream System

An aquifer-stream system is one where there is easy interaction between the stream and the aquifer. In normal circumstances, in a humid environment groundwater moves from aquifer to stream, so the stream is steadily gaining water and is sustained from the same source during dry weather. If, instead, it is an arid environment, the water level in the aquifer will be below the bed of the stream. Water will move from the stream into the aquifer and will steadily lose water as a result. In very dry times it may temporarily dry up. In both of these situations, humid or arid environments, the movement of water in either direction is dependent on the permeability of the streambed and stream banks.

The Arkansas River aquifer-stream system extends from the Rocky Mountains of Colorado to where the river joins the Mississippi in eastern Arkansas. Thus the system traverses semiarid, semihumid, and humid areas along its course. The river is thus affected by a range of climatic and hydrologic environments. In eastern Colorado the natural flow of the river is derived from snowmelt from the mountain areas. So while water is abundant, the flow is from river to aquifer; and in the same location, when there is drought, the aquifer is fed from the river.

In western Kansas the valley of the Arkansas River is underlain by alluvium, which is mostly composed of sand and gravel, and as it crosses the High Plains Aquifer system, there is a tight linkage between the two, creating one integrated aquifer system. For some distance inside the Kansas border, the High Plains Aquifer discharges into the Arkansas River, a result of the general slope of the land, providing most of the base flow of the river. In extremely dry weather, however, the river receives no water from the aquifer, and before long, it dries up. Perhaps because of the high rates of withdrawals from the High Plains Aquifer, the Arkansas River now rarely receives any water from this aquifer for a 24-kilometer stretch of Kansas territory.

YELLOWSTONE

Yellowstone National Park is often referred to as the nation's jewel of the park system. It was the first to be established. That was in 1872. It was also the world's first national park. It is perched on a series of volcanic plateaus in the northwest corner of Wyoming. Average elevation is about 2,400 meters. Because of its elevation and northern latitude, the climate is cold and humid. The center of interest in the park is the collection of hot emissions from deep inside the earth, geysers, mud pots, hot springs, and fumaroles. They constitute one of the greatest number of geothermal features to be found anywhere.

Traditionally, a geyser in Yellowstone, such as Old Faithful, would thrust its jet of boiling water high into the air about once every hour, and this was always a major attraction for visitors. Then in 1959, an earthquake at Hebgen Lake, about 60 kilometers northwest of Yellowstone Park, changed all that. The patterns of eruption of many of Yellowstone's geysers were permanently altered. Now they are less frequent. Such are the unpredictable nature of volcanic activities. Over geological time they are even less predictable. Ten million years ago, there was no volcano here, and 10 million years hence there will be nothing here to indicate its former presence.

The entire Yellowstone Basin is a volcanic caldera, what is left of a gigantic eruption that occurred in the distant past. Geologists estimate that an eruption of this kind happens approximately every 600,000 years. The last two were 1.2 million and 600,000 years ago, the last one having ejected 1,000 cubic kilometers of ash when it erupted. That would be more than a thousand times the amount of ash thrown up by Mount St. Helens in 1980. By all available evidence, this area is due for another massive explosion like the previous two, sometime within the next few thousand years!

Beneath ground level at some depth is the magma pool, and above it in the rock layers closer to the surface are numerous fault lines. Water from rain and meltwater seep down into these cracks. It expands as it gets superheated and is then forced up to the surface. Depending on the shape and size of these underground faults the superheated water takes different forms: It may be a fumarole, that is, a jet of steam, or a hot spring, which is a boiling pool of water, or it could be a mixture of steam and hot water, known as a geyser, an eruption that occurs in fits and starts under the influence of different types of pressures from below.

There are many other variations in the activities taking place within the caldera. There are mud pots, which are acidic hot springs that hold large amounts of minerals in suspension underground because they cannot flush away the minerals before they reach the surface. As a result they appear as muddy waters, with their bubbles making a glop-glop sound. Hence the name mud pot. Mammoth Hot Springs rises up through rocks that contain easily dissolved minerals. One of these minerals is calcium carbonate, and as it precipitates out of the hot water, on cooling it forms terraces around the mouth of this particular hot spring. All of these volcanic activities occur inside the 50 by 75 kilometer surface area of the caldera.

Yellowstone is a very significant place, unique in the coterminous United States, for quite a different reason than its volcanic history and present state. It represents what has come to be called a hot spot. We encountered this term and saw its effects in Hawaii where the long chain of former volcanoes across the Pacific provided evidence of a hot spot's existence. Over the past few decades it became clear to geologists that it is almost impossible to fix permanently, say, the latitude and longitude, of a place on the surface of the earth. Everything seems to be in motion, and the only thing we seem to be able to say is that such and such a place is fixed relative to some other. Plate tectonics revealed all this.

Hot spots are the one exception. Over millions and millions of years they preserve their locations with respect to the deeper part of the earth's mantle. Continents and ocean plates move over them and leave them where they are. In the case of Hawaii, which is located over a hot spot, we know from the chain of undersea mountains that formerly were volcanoes that the Pacific plate has been moving over it for the past 70 million years. It was a similar story with Yellowstone, but the evidence is not nearly as clear as in Hawaii. Continental movements are far more complex than ocean ones because the crust is far thicker.

The theory behind hot spots is that they are the surface expressions of what are called mantle plumes. Heat is continually radiating out from the earth's core into the deeper parts of the mantle, where, though solid because of the overlying weight of rock, some movement is possible. Over long periods of time, masses of rock deep in the mantle become so hot that they rise close to the surface. Heat expands them, makes them lighter, and therefore able to move higher within the mantle. One of these masses, it has been estimated, could have a volume of millions of cubic kilometers, enough to maintain a link with the deeper parts of the mantle and, at the same time, provide magma for a hot spot over millions and millions of years.

Searching for evidence that demonstrates Yellowstone's hot spot history is difficult. Many scholars are working on it. About 160 kilometers to the southwest, lava 6 million years old was located. This fits what we know about the direction of movement of the North American plate. It moves three centimeters a year toward the southwest. Farther to the southwest in Idaho, lava aged 10 million years was located, and still later, more than 600 kilometers to the southwest, lava flows 13 million years old were identified. These locations and times are in the right direction and very close to what would be expected. Beyond these places, tracings have been few. In Oregon, lava 60 million years old was found. The latest finding is indicative of the whole problem of tracing hot spot origins in continents. A geologist is convinced that the 70-million-year-old lava found on the borders of Alaska was once over Yellowstone.

Yellowstone is still very active. There is a high rate of heat coming from the various vents, much higher than from other parts of the United States. In total, if it were converted into electricity, there would be enough to supply the power needs of a 5 million population city. The question has to be asked: Why is this not developed? Would it not be good for the environment if it replaced an equivalent amount of coal? What are the downsides to transforming Yellowstone into an electrical generating station? Other countries like New Zealand and Iceland make good use of geothermal heat.

There is another side to the story about large quantities of heat. The heat source is shallow, no more than a few kilometers beneath the surface. About 30 years ago it was found that the floor of the caldera was rising. Then less than 10 years later it was found to be falling. These movements were accompanied by swarms of earthquakes, just as we observed in Mammoth Mountain of the Basin and Range Province. All these are indicators of seismic activity. The history of eruptions here, however, does not suggest imminent volcanic action, but no one is taking that for granted. Studies and measurements go on continually.

WATER

High Plains Aquifer

The High Plains Aquifer underlies parts of almost all states in the Great Plains, including Colorado, Nebraska, Kansas, New Mexico, Oklahoma, South

Figure 9.3. Teton Dam failure, Idaho. The Teton reservoir behind a 43-meter-high earth dam was being filled for the first time in the spring of 1976. On 5 June of that same year a leak developed early in the day. By noon the dam had collapsed, and water cascaded into the canyon of the Teton River. Eleven lives were lost, water spread over an area of 460 square kilometers, and damage amounted to $400 million.

Dakota, Texas, and Wyoming, an area in excess of 620,000 square kilometers. About 30 percent of all the groundwater used for irrigation throughout the nation is drawn from this one aquifer in order to meet the needs of agriculture. When there are concerns about declining water levels, it is time to monitor the aquifer and examine afresh what adjustments need to be made.

In the decades following World War II, irrigated agriculture made great strides in the Great Plains, and groundwater levels declined by as much as 35 meters in some places within a period of 30 years. The U.S. Geological Survey decided to conduct an investigation of the problem, and results were published in the early 1990s. Some 6,000 wells were measured between 1980 and 1993 and decreases in water levels documented for each part of the whole aquifer. Most of the biggest losses occurred in the central and southern areas (Figures 9.3 and 9.4).

Data from the USGS investigation show that water levels decreased more slowly in the late 1980s than they did before 1980 even though the acreage of irrigated agriculture had doubled. One reason for this is a well-known charac-

Figure 9.4. Damage from the Teton Dam failure. Floodwaters emerged from the canyon 8 kilometers downstream to flood the communities of Wilford and Sugar City with a wall of water three meters high. Other communities were equally devastated. The velocity of the water was very great, so the destructive power was correspondingly great. Not until the water was absorbed by the American Falls Reservoir, 160 kilometers downstream, did the rampage stop.

teristic of climate in this part of the country, the unpredictability of rainfall from year to year. In his book *Grapes of Wrath*, John Steinbeck describes the terrible conditions in the Great Plains when rainfall was at an all-time low and dust storms rather than crops were the lot of farmers. The opposite was the case in the years 1981 to 1992: Rainfall was almost a centimeter above average, and much of the aquifer was recharged.

Other developments in the 1980s and early 1990s further reduced the depletion rates. New irrigated lands shifted from the drier southern lands to the north where less water was required and where there were higher levels of recharge. Advances in irrigation technology, using center-pivot sprinklers and pipes designed to apply water more evenly, also helped to cut back on demand. Changes in crops were yet another development that helped to conserve water; plant varieties requiring less water were introduced. Finally, new powers were given to local authorities to regulate the amounts of water taken from wells.

Figure 9.5. A salt plain on the Cimarron River in the northwest area of Oklahoma. Here large deposits of salt were laid down in Permian times, created by evaporation of seawater. The salt is dissolved underground, then, aided by openings in the surface, is brought above ground and deposited in layers as shown here.

Salinity

Salt deposited in an inland sea in Permian times occurs in several thick salt units underlying large areas of Colorado, Kansas, New Mexico, Texas, and Oklahoma. The salt was deposited from evaporating seawater over 200 million years ago. Several of these salt units at shallow depths are now being dissolved by groundwater, and the resulting brine moves laterally and upward under hydrostatic pressure until it reaches the land surface where it forms salt plains (Figure 9.5). The following four conditions for this to be an ongoing process are present, so the problem is a continuing one: (1) salt deposits; (2) supply of water not saturated with salt; (3) access to the surface for brine; (4) source of energy to move freshwater through the system.

Brines are deposited in the Arkansas and Red Rivers and their tributaries from their sources onward, so their waters are useless for industrial, municipal, or irrigation purposes. Farther downstream, away from the sources of brine in such locations as south and east of Oklahoma City, freshwater inputs from tributaries dilute the brine sufficiently to make it usable. It takes some distance from the source for this to happen. The amount of salt fed into the rivers amounts to thousands of tons daily, and the underlying beds from which it is dissolved can be as thick as 100 meters.

The whole process of salt dissolution is self-perpetuating. As the salt is taken away, open spaces are created underground rather like the action we saw in the karst landforms of the Appalachians. Once the ground is sufficiently weakened by this undermining, it collapses. The opening up of the surface then facilitates the entry of freshwater, either from precipitation or surface drainage, and thus an acceleration in the rate of salt dissolution as additional quantities of water attack the salt.

Underground action on carbonate rocks takes place in all parts of the country at different rates, depending on the type of rock and volume and speed of groundwater. The resultant surface of the land assumes a characteristic pattern, often marked by sinkholes or depressions. The shape of the land surface above these salt deposits is unlike any of those shaped by carbonate dissolution because the rate of change is so much greater with salt. The surface of the land around the Red River can only be described as chaotic because of the collapse of rocks into small and large caverns. Beneath the surface the strikes and dips of strata are also chaotic, changing sharply within short distances. Frequently, small-diameter chimneys extend vertically through hundreds of meters of rock from deep dissolution cavities.

Edwards Aquifer

South and east of the High Plains Aquifer is another one, less extensive than the High Plains but one of the most prolific aquifers anywhere in North America. Two million people are served from Edwards Aquifer, including the cities of San Antonio and Austin as well as numerous smaller municipalities plus agriculture and industry. This aquifer receives 80 percent of its recharge from surface streams. Most of the rest comes from precipitation, but in this part of the nation, it is not so much the quantities of rain that fall so much as the unreliability of the supply that causes concern.

Over the 50 years 1940 to 1990 the recharge rate fluctuated wildly. In 1940 it was 12 cubic meters per second; in 1956, it was 5 cubic meters per second; and then at various times since 1956, the rates have ranged all the way from 10 to 40 cubic meters per second. The historical low of 1956 was sufficient to meet needs at that time, but it is quite inadequate for present needs. The area between San Antonio and Austin is one of the fastest-growing urban corridors in all of the United States. Here at times in the late 1990s undeveloped land was being subdivided for homes and businesses at the rate of one acre every three hours.

Texas has an unusual system of water law. Surface flows are owned and regulated by the state, but groundwater belongs to the landowner, who is free to use as much as he or she likes as long as the water is being used beneficially and is not being withdrawn negligently in a way that causes subsidence. The result of all this is that water has been pumped by landowners in times of drought with the result that members of endangered species were at risk of extinction and the people responsible were liable to prosecution under the Endangered Species Act. To forestall irreparable damage of this kind, the Edwards Aquifer Authority was established in 1993. It has power to regulate groundwater withdrawals.

Conflicts persist, no matter what regulations are introduced. Population growth intensifies water demands, yet there are no alternative water resources of the quality and cost available from the Edwards Aquifer. Attempts to store water in times of plenty are frustrated by the lack of suitable reservoir sites. Even if there were such, there are downstream cities like Corpus Christi that would object to any interference with their water supplies from the same source. There is still another objection to any interference with the natural flow of this groundwater: Wetlands along the Texas coast depend on the same water supplies as do the cities.

Houston

A case study of Houston provides a useful picture of the interconnectedness of geological faults, groundwater, and land subsidence (Figure 9.6). More than 80 active faults are known to exist within the Houston metropolitan area, and over the past decades, they have caused considerable damage to the buildings and highways that were built over them. Careful examination of the places experiencing damage indicates that most movements of faults occur where groundwater withdrawals are highest.

Apparently, fresh activity occurs in these faults as groundwater is withdrawn. The slight movement along a fault line as water is removed from the surrounding area and consolidation takes place is sufficient to cause considerable damage at the surface. This phenomenon was discovered almost by accident. Considerable land subsidence had been taking place along Galveston Bay and the Houston Ship Canal, so groundwater withdrawals in nearby areas were cut back in the 1970s to slow this damage to the coast. Soon after, it was observed that less and less faulting was occurring in the locations where groundwater withdrawals had been cut back, so the efforts to reduce subsidence at the coast provided the additional benefit of stopping the destructive faulting.

The Houston Metropolitan Area depends on groundwater for its municipal and industrial water supplies to a degree unmatched by any other U.S. urban area of similar size. As a result, groundwater levels dropped in the course of the past several decades, and these declines led to consolidation of the water-bearing sediments and hence land subsidence. A total area of more than 12,000 square kilometers has been affected by subsidence.

Historically the greatest depth of subsidence is three meters, but this has been

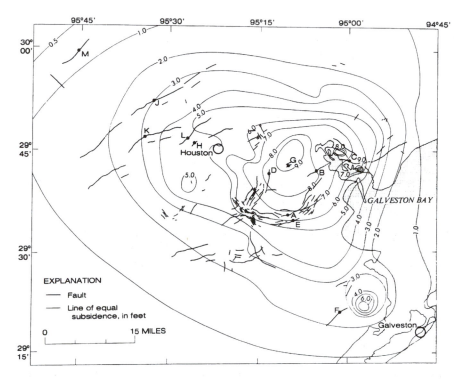

Figure 9.6. Map showing faults and amounts of land subsidence in the Houston Metropolitan Area for the period 1906 to 1978. Most of the fault slippages and consequent damage to buildings occurred where groundwater withdrawal was greatest. Areas close to Galveston experienced least damage because water withdrawals were at a minimum there.

enough to inundate permanently more than 10 square kilometers of coastal land in the vicinity of Galveston Bay and the Houston Ship Channel. Two dangers arise from this lowering of the land surface: (1) saltwater advances inland, and wetlands are abandoned by many forms of wildlife, as we saw under similar circumstances in the southern areas of the Coastal Plains; (2) the second danger relates to the heightened risk of tidal flooding whenever storms and hurricanes reach this coast (Figures 9.7–9.9).

The Future

The stresses over adequate groundwater supplies are likely to be greater in the future. Increasingly there are indicators of a rise in temperature and a drop in precipitation within the Great Plains. At first came the long-term indications, those that span a century or more in the life of the bigger glaciers in the north-

Figure 9.7. Concrete step-block seawall on North Padre Island, Texas, severely damaged by a storm surge during Hurricane Allen in 1980.

west of the region. Since 1850, the larger glaciers of Glacier National Park, Montana, have shrunk by two thirds. Some smaller ones have completely disappeared, and others have receded a kilometer and a half into the shade of steep cliffs. On the basis of present trends in global warming, it is expected that all the park's glaciers will be gone by 2030.

A long period like this of glacier change represents something more than weather cycles that come and go over 10- or 30-year time spans. These changes in Glacier Park are indications of a shift in climate, a change that will in time affect all parts of the ecosystem. Stream flow and temperatures in mountain watersheds will change, and plants, animals, and humans will have to adapt to the new regime. The evidence from the glaciers supports the convictions of climatologists that the earth's temperature has indeed been rising by at least half a degree Celsius over the past century.

Most models of climatic change predict an increase of average precipitation in winter at high latitudes because of the greater poleward transfer of moisture derived from increased evaporation at low latitudes. Tropical and subtropical land areas have seen precipitation drop over the past few decades. It is easy to overlook this development because these same areas get flooded when episodes of El Niño and La Niña occur. Increased intensity of rainfall is another outcome that accompanies higher temperatures, and this means greater soil erosion and higher risk of flooding. The Red River flood of 1997 was a shocking reminder of these anomalies.

Figure 9.8. Flooding in a low-lying area of Brownwood subdivision, Baytown, Texas, the result of surface subsidence, which, in turn, is due to extensive groundwater withdrawal.

RED RIVER FLOOD

We have already examined several examples of floods in eastern, central, and western parts of the country, some of them 100-year floods and some repeat 100-year floods even though the time between them in the same location was only a few years. The Red River flood of 1997 stands apart from all of these. It was a 500-year flood, and residents of North Dakota, Minnesota, and Manitoba in Canada will never forget their experience of it. Many cities were inundated in all three of these places. Once again we have to remember that these statistical statements about 100- or 500-year events may have exceptions. In the case of the Red, there was one in 1826 just as big as the one in 1997.

The Red River originates in Minnesota and flows northward through a large glacial lake basin, that of Lake Agassiz, the largest of the glacial lakes that were formed at the close of the last ice age. It covered an area of almost 1 million square kilometers and left in its wake a low-lying landscape, almost completely flat in places. The present extent of the Lake Agassiz Basin in the United States is 104,000 square kilometers. The Red River overflows its banks in most years, and the low elevation of the surrounding terrain ensures that water covers a big area when that happens. But the events of 1997 were much more, far worse than anything previously experienced.

The flatness of the river basin is evident in the gradient as the river flows northward, an average slope of 10 centimeters per kilometer for the whole 870

Figure 9.9. The Wink Sink in Winkler County, Texas. This collapse feature that formed on 3 June 1980 was caused by salt dissolution underground. As so often happens in events like these, this one expanded to 110 meters in width and 34 meters in depth within a period of 24 hours.

kilometers of its length. Natural levees, 1.5 meters in height, rise on either side of the river, so whenever these are overtopped, flooding spreads for several kilometers on both sides. In October and November of 1996 the first of several unusual conditions were noted: Rainfall was three times the average for that time, leading to total soil saturation. In such cases, all new rainfall becomes runoff. The second exceptional condition was snowfall through the winter, three times the average. Then in the spring there was another period of above-average rainfall.

Flooding began at the end of March and continued throughout April. The flood stage is 3 meters above base, but from the first day, water ran at twice that height. Community after community had to evacuate. By mid-April, flood-waters had reached a height of 12 meters. Grand Forks recorded a height of 16 meters about the same time. Damage was at a maximum in Canada because the province of Manitoba has almost three quarters of its population resident within the Red River Basin. By December 1997 the International Joint Commission, which coordinates relations between the two countries, met and mapped out a series of new preventive measures based on common experience from the previous April.

The Canadian federal government has already begun work on better dikes. Some $40 million has been allocated to build permanent dikes as protection for

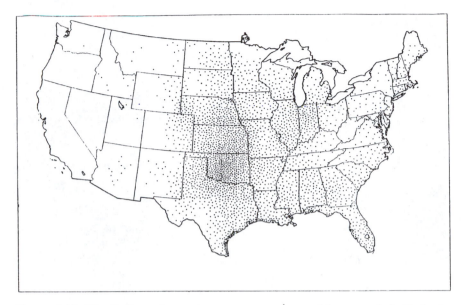

Figure 9.10. Distribution of tornadoes over an average 10-year period. Each dot represents two tornadoes. The main concentration is in the area we know as the Great Plains, but it is also obvious from the map that these very destructive storms are found all over the country, with a fairly high density in most states east of the Rockies.

11 of the communities that were affected by the 1997 floods. They are being constructed to a height of 60 centimeters above the highest level of the 1997 floods. The width of each dike will be big enough to drive a truck along it. The city of Winnipeg is already protected by a bypass floodway that was built about 40 years ago, but city authorities feel that the floodway has inadequate capacity and needs to be enlarged. The problem here is that this would force a number of farmers to sell their land to the city in order to make way for an enlarged floodway, and not one of these farmers is ready to sell. There will be some tense moments and a lot of bargaining before this difficulty is resolved.

TORNADOES

These cyclones, like hurricanes, form in warm climates, but the tornado is the most intense cyclone of all. Its most frequent and violent occurrences are found in the United States, mainly in the Great Plains Region and to a lesser extent in the Central Lowlands (Figure 9.10). April, May, and June seem to be the favorite times for strikes, with an average of more than 400 in the United States as a whole in these months. The famous *Wizard of Oz* movie was one of the first media events to raise awareness of a tornado's destructive power.

The tornado is a small dark funnel cloud, a few hundred meters in diameter at its base with a cumulonimbus cloud above it. Its dark color comes from the

dust it picks up by its powerful inward-spiraling winds. These wind speeds can reach 400 kilometers an hour as the storm twists and turns and races along the surface of the ground. It can devastate almost everything in its path, yet at other times it can rise in the air and leave the ground below completely unscathed.

The National Weather Service maintains a tornado forecasting and warning system similar to the one for hurricanes. Whenever weather conditions seem to favor tornado development, places at risk are warned and arrangements for observing and reporting conditions are set in motion. Western Ohio is not a high-risk location compared with states in the Great Plains, but on 4 April 1974, at Xenia, near Dayton, this state was hit with one of the worst tornadoes of the century.

It was not alone. Over the two days, 4 and 5 April, a rash of 148 tornadoes attacked 12 states, killing 300 people and injuring 6,000. At Xenia some 3,000 structures were demolished. Elsewhere, entire towns were wiped out, and $600 million worth of property was devastated. The atmospheric conditions were ideal for triggering tornadoes: A cool mass of humid air lay over Chicago, while farther west dry air was encountering a cold air mass from the northwest. Against both of these came a moist, warm air mass from the south. The combination of all three created an explosive series of thunderstorms extending more than a thousand kilometers from Texas to Illinois.

An earlier tornado swept through three states in 1925. It touched down first in Annapolis, Missouri, with a base at times as wide as one and a half kilometers. Main Street's buildings were flattened in a few seconds, and the twister then swept on into Murphysboro, Illinois, tossing trees, buildings, and even underground pipes as if they were toys. Over 230 people died in Murphysboro. The tornado moved next to DeSoto, a town of about 600, where it knocked down every structure more than one story high. Sixty-nine people lost their lives. The twister finally vanished in southern Indiana. Fortunately storms like this one or those of 1974 are rare but, sadly, not unique.

On the evening of 3 May 1999 the worst tornado of the century, as far as costs are concerned, touched down on Oklahoma City. It was the nation's first billion-dollar one. It was not alone. Other parts of Oklahoma, the state that gets more tornadoes per square kilometer than anywhere else on earth, were hit with 65 of these storms on that same evening, all of them in areas close to Oklahoma City. Within a period of five hours 8,000 buildings were in partial or total ruin as the rash of storms swept from southwest Oklahoma diagonally across the state toward Wichita, Kansas.

The difficulties involved in forecasting were evident on that fateful evening in May. The Storm Prediction Center (SPC), based at Norman, Oklahoma, issues bulletins every day, and on that morning's statement announced it as unlikely that any tornado would appear during the day. By early afternoon SPC raised its estimate to moderate. Not until close to 4:00 in the afternoon did SPC change its prediction to high risk—and then only because a powerful computer had shown that storms were charging across the state.

One hour later, across a 240-kilometer swath that included Oklahoma City, the swarm of storms struck. The greatest damage was caused in Oklahoma City and one or two of its suburbs. On the F-Scale of Tornado Strength, the one that hit the city was at 5, the top of the scale. Any F5 is unusual, and one that hits a major city even more rare. Street after street was razed to the ground. A typical sequence in a single-family home would be windows shattered, roof lifted off, walls caved in. Even homes that were carefully built to withstand 125-kilometer-an-hour winds were unable to withstand this tornado.

Mobile homes fared very poorly, as they usually do. An F1 tornado is usually enough to knock them over. With an F5 at speeds of 500 kilometers per hour, they were completely shattered. While the overall death toll was low for a storm of this size, most of the fatalities occurred in the mobile home areas. Almost every one of the tornadoes that hit over the six hours from 5:00 to 11:00 in the evening was an F5 or close to that strength. They were supercells, sustained severe thunderstorms, and experts were left with the problem of how such a powerful series of storms could be sustained at that level of strength for so long.

Most tornadoes develop within very large storms called supercells. These storms are found in unstable environments in which wind speeds vary with height and where cool, dry air rests on top of warm, moist air, with a thin, stable layer separating the two air masses, a condition similar to temperature inversion in other settings. If a weather system reaches this unstable mass, the status quo is disrupted: The low-level air is forced upward, and a vertical vortex gradually takes shape as the warm air ascends, cools to the point of condensation, then is triggered into faster ascent as the latest heat of condensation warms the surroundings.

Mitigating Damage

The problem of finding adequate, good-quality water for our cities and for agriculture is a national one but is especially challenging in areas like the Great Plains where total supplies are limited and climatic conditions for recharging aquifers unreliable. Authorities must therefore undertake to do one or more of the following: (1) find new supplies; (2) decrease demand; and (3) initiate better management and more efficient use of existing resources.

Tornado prediction is still far from being an exact science. In spite of their enormous power, they seem to thrive on a delicate balance of forces. Computer models of the future may be able to diagnose the preconditions for a tornado hours in advance of its striking. They will need to be fed much more information than is available at the present time if they are going to succeed. Dense surface networks are already in place across Oklahoma, but they need to be supplemented by more sophisticated radars. Portable radars of the kind needed are already under development in Norman, Oklahoma, at the National Severe Storms Laboratory.

REFERENCES FOR FURTHER STUDY

Church, C., and C. Nguyen, eds. *The Tornado: Its Structure, Dynamics, Prediction, and Hazards*. Washington, D.C.: American Geophysical Union, 1993.

Elias, Scott A. *The Ice-Age History of National Parks in the Rocky Mountains*. Washington, D.C.: Smithsonian Institution Press, 1996.

Fournier, R. O., et al. *A Field Trip Guide to Yellowstone National Park*. Washington, D.C.: U.S. Geological Survey Bulletin 2099, 1994.

Kessler, Edwin, ed. *Thunderstorm Morphology and Dynamics*. 2nd ed. Norman: University of Oklahoma Press, 1986.

Reader's Digest Association. *Marvels and Mysteries of the World Around Us*. Pleasantville, NY: Reader's Digest, 1972.

Saarinen, T. F. *Perception of Drought Hazard on the Great Plains*. Chicago: University of Chicago Press, 1966.

Smith, Peter J. *The Earth*. New York: Macmillan, 1986.

Sparks, R.S.J. *Volcanic Plumes*. New York: John Wiley and Sons, 1997.

10

CITIES' GEOLOGY

For most of recorded history, little thought was given to the rock structures beneath cities. New settlements were sited where they could easily be defended or where good transportation routes and adequate water supplies were readily available. Later, when a city emerged and problems arose with the underlying structure, it was too late to start all over again. The consequences of this tragic neglect can be found in many countries. The Leaning Tower of Pisa is one of the best-known examples.

This famous tower was started in the twelfth century but not completed until the fourteenth. Life was more leisurely in those days. If all of the base had settled evenly, all would have been well—and there was reason to believe that that would happen. The ground under the tower was 8.5 meters deep, consisting of sand, which was excellent foundation material. However, below the sand lay a stratum of salty clay of varying thickness and uncertain strength, and it was this material that gave way in places under pressure from the load at the surface and caused the tilt. Recent action, using heavy weights in some locations and removing soil from other places, are proving to be successful in preventing the tower from tilting any farther from the vertical.

Planning within cities must take account of the ground beneath, not just the surface layers, that is, the nature and layout of the ancient rock structures hundreds of meters below ground. This will include an understanding of the porosity of these rocks and the quantities of water stored in them. Some understanding of geology is essential, particularly the ability to read geological maps, if adequate use is to be made of available knowledge. Regrettably, none of this detail was attended to throughout most of the twentieth century, either in European or in North American settlements.

One notable exception is the city of Boston where geological foundations were more thoroughly investigated than in any other American city. From as early as 1914, records were kept of boreholes put down within the city, and in 1931 a local journal described plans to gather data on the character of subsoils throughout the Boston area in order to get a clear idea of the geological construction of the city.

All of the major North American cities were built with little regard to underlying geology. When the wake-up call came with the San Francisco earthquake of 1906, it was too late to do anything more than learn to cope. The need for attention to the ground beneath our cities is now greater than ever because the vast majority of Americans live in big cities, a fundamental shift from the early years of the twentieth century when most people were in rural settings. Any hazards that affect cities will now impinge on large numbers of people. The immediate need now is a better understanding of the geology of cities so that authorities will know how best to cope with problems as they arise. In this connection, locations of geological faults under cities is a high priority.

Los Angeles is one prime candidate for number of faults. No other city or urban complex has so many faults within its boundaries. They are both numerous and complex. Though they are fewer than in Los Angeles, many other cities of the Cordillera also are built on or near faults. In San Diego, there is a strike-slip fault like the San Andreas, the Rose Canyon Fault, running north and south and slipping at the rate of 1.5 millimeters per year. San Francisco is close to the big San Andreas Fault, but it has many others as well. East of San Francisco, in Oakland, there is yet another major fault, and it threatens several of the other cities in the general area of the Bay. All these Californian faults are examples of the kind that can be found all over the urban world, in Tokyo, Bandung, Athens, Lima, or Seattle, to mention just a few.

SAN DIEGO

Geological issues rose sharply in the early 1960s as more and more people came to live here and as subdivisions moved into steeper and steeper terrain. Most of the city still stands on a 40-kilometer coastal strip, underlain by relatively flat, westward-sloping sediments. Mount Soledad, a peak lying west of the Rose Canyon Fault, is an exception to this general pattern. The Rose Canyon Fault, the only one within the city, passes directly under the downtown, and while there has not been a damaging earthquake in historic times, it is inevitable that one will come sometime. While the built-up part of the city has only one fault, there are several others close enough to be matters of concern.

It is unfortunate that the importance of geological studies is rarely recognized until some destructive event draws attention to it. In the case of San Diego the event was a landslide on Mount Soledad in 1961 that destroyed nine homes. Four years later, another slide at a different location took out a further eight homes. Both sets of homes had recently been built on geologically hazardous ground, but no one had conducted a study of the area before construction began.

Soil investigations were mandatory, but they were useless for identifying seismic risk. When geologists studied Mount Soledad, they discovered new fault planes, and it was along one of these that the failure occurred, causing the landslide.

The 1965 landslide, it was discovered later, happened at the site of an ancient landslide, and this discovery led to intensive research into other possible ancient landslide sites within the San Diego urban area. When one of these locations is identified, steps are taken to stabilize the site. A much greater risk relates to homes on the cliffs that stretch for about 80 kilometers all the way to Orange County on the outskirts of Los Angeles. These cliffs vary in height from 5 to 70 meters, and they are highly desirable and valuable view sites for San Diego's residents. Geologic studies are required before any construction is undertaken.

These cliffs typically have a lower vertical part of older rocks, extending upward for 8 meters from the beach, and an upper area of more recent material that slopes toward the sea at an angle of 45 degrees. The average arrangement for homes is to set them back 18 meters from the cliff face. What is never known for sure is the rate at which cliff recession will occur. Landslides, block falls as waves undercut the upper layers, and the ongoing process of wind and water erosion all act to cut back the cliff at rates that can be predicted in the short term. Over longer periods, however, climate changes and unexpected earth shakings occur, accelerating or slowing down cliff recession. In one location, 200 meters of cliff were lost within eight years as a result of a series of unusually heavy storms.

LOS ANGELES

From the first days of the San Andreas Fault, following the contact of the North American plate with the East Pacific Rise, the area in and around Los Angeles has been taking shape from terranes. That period began 29 million years ago, and since then, various pieces of crust have been drifting northward to form part of California. Many millions of years later, the Los Angeles Basin received huge quantities of organic materials that, in modern times, became the source of today's oil. Early in the twentieth century, intensive oil production began in Los Angeles, and, before long, in many surrounding areas. Soon the problem of subsidence appeared.

In one oil field, in the early 1950s, subsidence rates of more than a half meter per year were observed. The cost of repairs mounted over the years; and when, in 1963, surface faulting occurred, leading to damage to oil wells and property, there was a public outcry. Subsidence was corrected by water injection, and this proved to be an effective and low-cost method. One rare oil seep at Rancho La Brea revealed much more than the presence of oil. Here were tar pits that had preserved the remains of 200 different kinds of mammals, birds, reptiles, insects, and plants. More was learned about Pleistocene life from this place than from any other single source.

Just as was the case in San Diego, rapid increases in population in the 1950s and 1960s forced development onto the higher ground as the lower levels became fully occupied. Hills and even mountains were shaped to fit the needs of the new subdivisions without much regard to the natural environment. Results were devastating: Homes and lives were lost in the landslides and debris flows that ensued. The government tried to correct the problem by issuing a grade code, reducing the degree of slope of all roads and highways, but the geological processes were not addressed, and conditions continued to get worse. Finally a new code was introduced, one that took account of both geological processes and engineering controls.

The large number of faults in the Los Angeles area poses its own set of problems for builders. The government prohibited the construction of buildings that straddled active faults. Fault study zones were identified throughout the city, within each of which seismic studies were required. Earlier codes provided for the elimination of parapets and the reinforcing of masonry structures in order to minimize the effects of severe earth shaking during an earthquake. The new code, based on recent studies of faults, requires builders to design homes and other buildings so that they will withstand a defined level of shaking without experiencing structural damage. The level required is defined by location with respect to particular faults.

SAN FRANCISCO

The San Andreas Fault is the dominant geological feature in all of California. It runs past the city of San Francisco; it continues northwestward to Cape Mendocino, where it meets the Juan de Fuca plate; and it extends southward to the U.S.–Mexican border near the head of the Gulf of California. It is the boundary between two huge tectonic plates, the North American and the Pacific, and is the main cause of the big earthquakes that both the city of San Francisco and the rest of California have experienced over the past two centuries. In addition, there are numerous faults associated with the main one, and these continue to trigger lesser shocks.

Earthquakes are not the only geological hazard in and around the city of San Francisco. Landslides are common wherever there are steep slopes and weak soil surfaces. Thousands of landslides follow earthquakes. The famous 1906 quake triggered more than 10,000 landslides over an area of 32,000 square kilometers. Three thousand occurred as a result of the 1989 Loma Prieta quake near Santa Cruz, about 80 kilometers south of San Francisco. Intense rainstorms also cause debris flows. In 1982 San Francisco received 616 millimeters of rain in 34 hours, and the result was a devastating series of 18,000 debris flows, killing 25 people and causing $25 million worth of damage. A few years later, 800 millimeters of rain fell over a period of nine days onto a land surface that was already saturated by 400 millimeters of prior rainfall. Again, extensive property damage was the result.

Despite these abnormally high rainfalls, California is frequently short of adequate water supplies, and groundwater has to be tapped to supplement surface flows. Over a 50-year period in the first half of the twentieth century, years and years of deficient rainfall led to a fourfold increase in rates of withdrawal of groundwater in the clay-rich beds beneath the alluvial lowlands in the southern part of San Francisco Bay. As a result the ground subsided as much as four meters in the San Jose area as pressure decreased in the underground aquifers, allowing adjacent clays to collapse. Saltwater then flooded the area, and the foundations of several buildings were damaged. Since that time, increase of surface water imports and new controls over withdrawals from groundwater ensure that these subsidences will not recur.

Extensive research endeavors are being conducted throughout San Francisco to gain a better understanding of earthquake risks and, at the same time, to set standards for building codes in places where there are active faults. The city has eight major faults, and these are constantly being monitored by networks of instruments to identify the active ones. Where buildings are being erected near an active fault, developers are required by law to employ a registered geologist to advise on danger spots. Typical fieldwork by the geologist includes excavating trenches across fault lines and measuring slippage over the previous 10,000 years. One mistake from building practice at the beginning of the twentieth century will not be repeated: erecting homes on landfill sites near the water. During the one minute of shaking in the 1906 quake, all these homes collapsed.

While the peninsula on which San Francisco stands is dominated by the San Andreas Fault, the other side of the Bay, in Oakland, is focused on the Hayward Fault, one of the most studied anywhere in the world. This fault is active, it runs through a densely populated area, and it has caused major earthquakes in the past. It is a strike-slip fault, just like the San Andreas, and it trends in a similar northwesterly direction. Its length is 280 kilometers. If a rate of six millimeters per year is assumed, then the recurrence interval for the Hayward is 87 years, and experts have added to this a 96 percent probability that a quake of 6.5 or more will occur before 2032.

RENO

Reno is on the western edge of the Basin and Range Province, and the area around it forms a basin between the Sierra Nevada and the Virginia Range, a volcanic zone to the east. The Comstock gold and silver mining district is in the Virginia Range, and its fame in the middle of the nineteenth century is closely tied to Reno's history. As a basin the environment of Reno is covered with thick layers of recent materials, much of it outwash sands and clays from the last ice age. The Truckee River, which comes from Lake Tahoe, flows through Reno in a broad trough and continues eastward through a narrow canyon in the Virginia Range. Because of its location, earthquake hazards are evident throughout the city environs.

The Sierra Nevada mountains just west of Reno have a number of north-south trending faults that separate them from the Basin and Range Province. These faults are numerous and complex. They extend for more than 100 kilometers southward into California. As is the case with so many parts of the Cordillera, this region experiences both dip-slip and strike-slip fault movements, probably related to some extent to the larger strike-slip activity of the San Andreas Fault. Although no very large quakes have hit Reno in historical times, the potential for one of 7.5 strength is always there.

Various attempts are being made to estimate the recurrent time for such a big earthquake. Moderate quakes hit Reno in the 1860s and 1950s, typical for much of the Basin and Range Province, and therefore there are expectations that a quake greater than magnitude 7 could happen every 40 years. For more powerful earthquakes, the intervals might well be in thousands of years, enough to lull authorities into assumptions that nothing serious will happen in the twenty-first century. We know from experience elsewhere that this is not responsible think- ing. The longer the estimated time interval, the less the available data—and hence the unpredictability of an event.

The threat of landslides is never far away in a mountain zone such as this one, especially since there are major faults here as well as jointed granitic rocks. About 25 kilometers south of Reno stand Mt. Rose and Slide Mountain, both with much of their surfaces of the jointed granitic type that is easily eroded, especially during heavy falls of rain or snow. In the winter of 1982 a large mass of rock broke off from Slide Mountain. There had been high volumes of pre- cipitation for some time, and this was followed by a warm spell in late spring. The volume of rock and debris that came down was more than 700,000 cubic meters. It was an avalanche rather than a landslide, and it left heaps of material in several places ranging in depth from 10 to 30 meters. Stream channels were either blocked or deeply eroded by the debris, and several homes were de- stroyed.

SALT LAKE CITY

This city is the largest between San Francisco and Denver, and it serves as a regional center for five states. Like Reno it is surrounded by mountains and has a river, the Jordan, running through the city. Earthquakes are common along a line that stretches from Arizona, through Salt Lake City, to Montana, and they are often of magnitude 7 or more. Two from the second half of the twentieth century are particularly notable, Hebgen Lake, Montana, in 1959, magnitude 7.5, and Borah Peak, Idaho, in 1983, magnitude 7.3. Earthquakes here are shal- low, and that means more destructive than deep ones, but their epicenters do not seem to coincide with any known geological structures.

The most obvious threat to this city is the Wasatch Fault at the base of the Wasatch Range, a 36-kilometer section of which passes through some of the more densely populated parts of Salt Lake City. There is evidence of past quakes of magnitude 7.5 or more, and these have included surface faulting on more than one occasion. Some of these past earthquakes occurred at intervals of more than 1,000 years, and this may lull present authorities into ignoring seismic risks. It is easy to say now that it would have been good if the early settlers had a geologist in their party, but perhaps it might not have made any difference. As recently as the 1970s, with full knowledge of the Wasatch Fault and its location, public authorities in Salt Lake City were willing to allow construction directly on top of this fault. Only strong public protest convinced them to stop.

Although there have not been any major quakes within historical times, the threat from an earthquake on the Wasatch Fault, whenever it might come, is very great. Hebgen Lake in 1959 and Borah Peak in 1983, both in the same seismic zone as Salt Lake City, are reminders that not only ground shaking and surface faulting can happen, but block subsidence also takes place. If the valley area in which Salt Lake City sits were to tilt slightly toward the east along the fault line, the city could be flooded from the Great Salt Lake. One other thing needs to be mentioned: There are extensive areas of unconsolidated sediments in and around the city, and we know from experiences elsewhere that this kind of terrain experiences liquefaction when strong earthquakes come.

By virtue of its mountainous setting, the city is always at risk from landslides and floods, especially when precipitation levels are unusually high. In anticipation of this possibility, some of River Jordan's tributaries had been diverted into storm drains under the city. In 1983, these drains were unable to carry the water that followed the heavy rains of that year, and the city was flooded. A number of corrective measures were taken afterward to forestall any recurrence of flooding: Sediment basins were built, stream banks were built up, and stream channels were dredged. One of the unusual threats of flooding comes from the Great Salt Lake: Because of its flat bed, the lake level is directly affected by rain; in the 1980s a period of heavy rain caused the lake to overflow and flood many valuable properties.

DENVER

Denver straddles two major geological regions: the igneous and metamorphic foothills of the Front Range to its west and the Piedmont part of the Great Plains on its east (Figure 10.1). Like San Francisco, it is a relatively new city, and in the last 30 years of the twentieth century, it has grown so rapidly that geological hazards assume new proportions. Not least among these hazards are the problems of waste disposal, including radioactive wastes, problems that demand action because of careless practices in the past. This city is both the capital of Colorado and the main commercial center for a string of cities stretching along the Front Range from Fort Collins in the north to Pueblo, 200 kilometers to the south.

There are four main hazards within the city limits, all directly related to geology. The most widespread and costliest is the problem of swelling soil and rock, which affects about half of the urban area. A further problem is the col-

Figure 10.1. Denver. The city is built on a boundary zone between the igneous and metamorphic foothills of the Front Range and the Piedmont part of the Great Plains. Mass movements of soil in the steep terrain to the west are common environmental problems.

lapsible and compressible soils found in about a quarter of the built-up land. There is a third danger, that of mass movements of soil, in the steep mountainous territory to the west. Finally, ground subsidence occurs in a number of locations.

Expansive soils have caused damage to buildings and roads. These soils swell when the local clay minerals get wet. Their crystal structure allows them to absorb water. They shrink when dry. Denver's downtown skyscrapers have taken account of this problem, and so their foundations are designed to withstand any uplift from swelling. Schools and single-family homes are not so fortunate. Their lightly loaded foundations cannot cope with soil movements. One state school had to spend one third of the cost of the original building on repairs to cracked walls, floors, and ceilings. Heaved and cracked city streets and highways are familiar sights in and around the city.

In Denver's eastern and southeastern sections, collapse-prone soils are found. They are low-density sandy soils that can carry heavy loads when dry but lose their strength and settle or collapse by as much as 15 percent when wet. Single-family homes again are the principal losers. Foundations move and structural damage follows. The third geological problem affecting Denver is concentrated on the west side on the steep slopes of the foothills and mountains. Vegetation cover is sparse, so slope failure, rock falls, and debris slides are common experiences after heavy rainfall.

The fourth hazard is a pattern of subsidence problems. One part of this is a series of old sand and gravel pits that subsequently were used as waste-disposal sites. Over time, organic material in these sites decomposed, sank, and produced methane gas. This double hazard caused explosions in buildings that were later constructed on top of these pits. Even buildings that were tied to foundations below the waste materials experienced rupture of water and sewage lines. Additional problems of subsidence came from underground coal mines that were worked in the early 1900s and over which buildings were later constructed.

KANSAS CITY

One feature of Salt Lake City that I did not mention is its use of underground storage. Two storage vaults were excavated in Little Cottonwood Canyon, extending 200 meters into the mountain. Inside these vaults, complete business activities were conducted. Water was easily collected at the back of the tunnels because of the widespread shallow distribution of groundwater. I mention these things here because Kansas City has become known as the world's leader in secondary subsurface space utilization. Warehousing, controlled temperature food storage, manufacturing, and offices have operated in Kansas City's below-the-surface world.

Kansas City, on the border between Missouri and Kansas and on the Missouri River, is in a region of simple geologic structure, in contrast to the complex systems of the Cordillera. It lies in a belt of nearly horizontal ancient rocks that dip toward the northwest at 3 meters per kilometer. These rocks are mostly shales and limestones and in the vicinity of Kansas City are 300 meters thick. The more recent surface deposits consist of dissected glacial till, including layers of loess. Like any other city, this one has its share of environmental problems, but they are masked by the imaginative development of underground areas.

At an earlier stage, Kansas City was extensively exploited for its limestone rock, a key resource for highway construction. At first it was mined by opencast methods, but as the overburden became thicker and thicker, the work went underground. The spaces thus created, and which continue to expand as more and more limestone is extracted, have now become the center of the city's unique industry. Mining practice here was and is quite enlightened, unlike what has often been described as the rob-and-run habits of some mining companies elsewhere. Substantial roof thickness is maintained, and the portions left as support pillars are generous. Limestone layers are almost flat, so both roofs and floors are level and smooth. Furthermore, beds of impervious shales successfully keep out water.

Development of the underground space began in the 1950s. At first it was used for vehicle storage, then for other purposes. The extremes of weather that characterize this part of the nation can be reduced or even eliminated below ground, and this element alone is essential for many products. Kansas City is ideally located as a place where goods can be temporarily stored en route to western destinations. Thousands of workers and many millions of square meters of used space can be found here at any time. New areas come into use as more limestone is extracted. Overall these enterprises are located on one level, at depths ranging from 10 to 50 meters below the surface. Construction costs are much lower in these underground spaces than they would be on the surface, and the heating costs are far less, too. During the cold war era, when fears of nuclear attacks were strong, Kansas City's underground was identified as a valuable asset in case of an emergency.

NEW ORLEANS

Like Chicago in its early battles with flooding, New Orleans' principal concerns center on water but to a far greater extent than Chicago ever experienced. It might even be said that the greatest engineering challenge facing this city at the mouth of the Mississippi is to keep it from drowning. Additionally, the city has very weak conditions in its foundation, so much so that it has been described as the flattest, lowest, and geologically youngest of any major city in the United States. The average elevation is 0.4 meters above sea level, and no surficial deposits are older than 2,500 years. About half of the urbanized area is at or below sea level. Floods on the Mississippi at times reach 6.5 meters above sea level, and hurricane surges on Lake Pontchartrain to the north of the city have exceeded 2 meters above sea level. Rainfalls of 250 millimeters within a period of 12 hours have been recorded on several occasions.

It is rare to find a city whose unconsolidated foundations date within the period of human history. They belong to the Holocene epoch and range in depth from 5 to more than 15 meters. New Orleans is about 75 kilometers from the Gulf of Mexico and more than twice that distance from the mouth of the Mississippi. It is part of that river's delta, a broad region of bayous and wetlands. Nowadays the main built-up part of the city is free from marshes as a result of the extensive measures taken to drain or pump away the water. Both natural and built levees run east and west within the city between the Mississippi River and Lake Pontchartrain, which stretches northward for more than 30 kilometers.

Levees extend along both sides of the lower Mississippi for a total distance of 2,500 kilometers. Farther up the valley of the river these levees are quite high, as much as 12 meters with base widths of 120 meters, but those around the city area average only 5 meters above the natural levee ridges on which they were built. Because the differences in elevation between the water level inside the levees and the lowest parts of the city are so big, there is a great need for a thoroughly dependable levee system. Fortunately the natural levees overlie coarse-grained inorganic deposits, and these are the best shallow foundation soils in the New Orleans area. In addition to the levees, there is a floodway through which water can be bypassed during a river flood.

As far as the city is concerned, hurricane-induced flooding can be just as catastrophic as a Mississippi flood. Rarely does a hurricane pass over the center of New Orleans, but when it happens, the devastation is widespread and costs are enormous. Flooding of populated areas is a certainty. The amount of advance warning is usually less than a day because, although its path can be traced for several days before it strikes land, its behavior as it approaches landfall is unpredictable. What can be done when flooding occurs? To move even a small percentage of the city's population to safe ground out of town cannot be done in a day. The only practicable alternative is to evacuate vertically, that is, move people to floors of homes or buildings that are above flood level.

Diversion of floodwaters is the usual method of minimizing threats to the city. To the west is a large floodway beginning far upstream and continuing down the Atchafalaya Basin into the Gulf, affectionately called the Old River Control Structure. Half of all the water in the Mississippi when it is at flood stage can be carried by this bypass. On the western outskirts of the city, on the main river, is another diversion, the Bonnet Carre Spillway. It can be opened to divert water from the river into Lake Pontchartrain. It is seldom used, but it is always available. There is a continuing concern about the stability of these protective measures because of the nature of the underlying sediments. During a major flood in 1973, for instance, part of the Atchafalaya was undermined and one wall failed.

Because there is so much unconsolidated material everywhere in and around the city, compaction of these sediments from time to time is the major cause of subsidence. Land sinking, shoreline erosion, and saltwater encroachment all are active and add to this problem of maintaining a consistent level of land. At times, these forces cause sudden changes to buildings and facilities. Differential settlement, bank failures, and flooding are the sorts of things that happen. If allowance is made for sea level variations, the general picture of subsidence rates is about 175 millimeters per century. Local groundwater withdrawals further aggravate the situation.

About 1 in 10 homes and the same proportion of commercial buildings, plus one out of every three streets and sidewalks, show signs of differential subsidence. Structures on the natural levees rarely are at risk, but the large number built on organic swamp and marsh deposits stand on a very unstable base. Typical conditions include buckling of patios and exposure of foundation slabs. Driveways, too, subside to such an extent that it is impossible to drive into carports. Gas and water leaks occur as underground utility lines sag. The problem worsens with development as new impermeable coverings of streets, parking lots, and buildings lead to dewatering and compaction in the organic soils beneath and hence subsidence.

When the first settlers occupied some high ground on the banks of the Mississippi almost 300 years ago, there was little thought about the problems of growth, but gradually, as the settlement expanded, the risks increased. Today the city continues to push its frontiers farther and farther into low-lying marshy tracts where building is possible only with the best of modern technology. Structures 200 meters tall stand where formerly the ground could not support the weight of one person. Water levels, when the river is in flood, can be as high as nine meters above the lowest areas of the city. There seems to be great faith in the stability of the dikes, but those responsible for them are always on alert, especially when strong winds blow.

Early building techniques used the natural levees. Crossed timber supports and masonry footings constituted the foundations. Later, piles were introduced for the bigger structures. These piles were driven down to the first sand stratum at a fairly shallow depth, where sufficient resistance was encountered to indicate a safe foundation. In the late 1930s, one 20-story hospital was constructed in this way, with piles that went down seven meters, but within a year or two the building began to settle, and before long, it had to be abandoned. Unstable layers of deposits beneath the sandy foundation gave way. At that time there was little detailed knowledge of subsurface geology, so no one knew about this weakness.

Over time, thousands of borings to depths of 60 meters or more have identified the nature of the underlying layers, not only the sand strata that seemed to be strong enough to hold up buildings but beyond that into the deeper Pleistocene deposits. When, in the late 1950s, a second hospital was built close to the site of the former failed one, over 2,000 piles were driven 25 meters into the ground, deep enough to reach the Pleistocene deposits even though the building had only nine stories. Some settlement of the ground was anticipated, and construction plans took account of this. That building has stood well, evidence that local authorities learned their lessons from the failure of the previous hospital.

The Pleistocene deposits are now the bedrock on which buildings need to rest. Where they are close to the surface, pile lengths and numbers can be few. Even so, there are deeper strata within the Pleistocene where compaction occurs if the load is great enough. The general rule now is this: The higher the building, the deeper the piles. Both the number and type of concrete piles are other considerations. A 1968 building of 45 stories had piles going down 50 meters, and a still more recent one, having 50 stories, used octagonal piles with diameters of 50 centimeters and depths of 64 meters.

CHICAGO

From its early days as a city, the flat terrain in and around Chicago created an ongoing problem of flooding; as settlement increased, the problem got worse (Figure 10.2). The land here is the result of the last phase of the ice ages, so there are end moraines and lake deposits, with all of them underlain by till. All of these glacial deposits are on top of an old bedrock of dolomite. One of the effects of glaciation is to disrupt long-standing watercourses, and this leads to swamps and marshy areas. To provide for new subdivisions, ditches were dug to drain the area, and water was obtained from private shallow wells. Sanitation consisted of outdoor privies, and under heavy rainfall, these privies often overflowed and contaminated the wells from which the drinking water came.

Figure 10.2. Chicago, looking north along the western shore of Lake Michigan. The challenge of flooding was a constant threat throughout the city's history because of low elevation and because of the disruption to historic drainage channels caused by the Pleistocene ice ages.

Throughout the second half of the nineteenth century, there were repeated problems with sanitation and water supplies in spite of the major efforts undertaken to remedy them. A water intake pipe was built out into Lake Michigan, and a canal for drainage and a water distribution system were installed in 1854, yet a heavy rainstorm overloaded the system, and the city was flooded. The resultant cholera epidemic killed more than 1,500. After this, water intakes were further expanded into the lake, and various improvements were made in the water and sewer systems. The canal was deepened, and large pumps were installed to increase flows. The work was interrupted by a fire in 1871, so these changes were not completed until 1881.

Looking back at these tragic events from today's standpoint raises questions about the wisdom of allowing so many people to settle in Chicago before basic services were installed. The city kept growing at a rapid pace. Its location was ideal as westward expansion exploded in the years following the Civil War, but the low gradients everywhere, the legacy of disrupted natural drainage from the last ice age, and a regime of moderately heavy rainfall made the challenge of providing good water and disposing of wastewater almost too much for the city authorities. Thirty years after the heavy rains and ensuing epidemic of 1854, a similar heavy rainfall overwhelmed the system, and cholera, typhoid, and dysentery claimed 175,000 lives.

In the face of this terrible tragedy, something that no other city in the United States had encountered, planners went back to the drawing boards to find a more permanent solution to recurring flooding and epidemics of disease. The bold new plan that emerged was to reverse the flows of local rivers and their tributaries so that they ran southward into the Mississippi River system rather than into Lake Michigan. Construction on the Chicago and Sanitary Ship Canal began in 1892 and was finished eight years later. In addition to reversing the flow of water, this new canal provided navigable access to the Great Lakes from the Mississippi River.

The city's sewers continued to empty into the rivers and canals, so water was drawn from Lake Michigan to dilute these wastes and flush them down the ship canal. Protests came from neighboring states and from Canada over this profligate use of lake water, and before long, the courts decided that Chicago would have to reduce its intake from the lake. In order to maintain sanitary conditions in the various waterways, the Chicago Sanitary District was forced to launch a massive construction project to build large-diameter sewers and sewage treatment plants. Not until the 1960s were the problems of freshwater supply and control of sanitation fully resolved. A system of tunnels and reservoirs was then initiated, and, with modifications, this system is still in place.

The tunnels and reservoirs plan provided for two systems: One captured and treated polluted sewer overflows, while the other retained excess storm waters. For the second system, which applied to the whole of the urban area, holding basins were built alongside streams to hold excess storm water until the stream could carry it. For the first system, which applied to the city of Chicago with its combined sanitary and storm sewers, vertical drop shafts were located to trap overflows from the existing system. Extensive tunnels were then excavated to convey the overflow water from the vertical shafts. Finally, pumping stations took the polluted water to waste treatment plants before returning it to the waterways.

All the efforts to find a working plan for Chicago's sewer and water systems produced a great understanding of local geology. It is likely that more is known now of the geology beneath Chicago than about any other city in the United States. The bedrock is ancient dolomite rock, hundreds of millions of years old, the surface of which is a rough plain with many deep valleys, the result of glacial scouring. These valleys are often as much as 30 meters deep, with steep, sloping sides, filled with glacial deposits. On top of the bedrock are the glacial materials from Pleistocene times, often 60 meters thick. Repeated fluctuations of glaciers caused different types and locations of the various deposits, and the high-level stages of Lake Michigan created its own landforms near its shores.

Detailed geological data on Chicago's underground proved to be of enormous value for local administrators, particularly in regard to zoning. Foundations for high-rise buildings have to be anchored in the dolomite bedrock that lies 30 meters below the surface, so the knowledge that steep-sided deep valleys cover that surface alerts engineers to the need to set foundations away from these slopes. Using open steel pipes similar to those employed in high-rise construction in other cities, very tall towers can be built safely. Potential areas of construction materials within municipal boundaries can be identified and protected until needed by the municipality for its construction projects.

Large cities like Chicago require more and more of their infrastructure to be underground. Chicago has the advantage of a uniform blue clay with little interference from faults, and this means greatly reduced costs for tunneling. For construction of roads, railroads, or airports, it is essential to know the nature of the underlying rock so that subsidence or tilting will not appear as a surprise years later. Equally important is a detailed acquaintance with the kinds of sand and gravel deposits that can be mined locally. Not every rock is suitable for road construction. There is one other concern that relates to construction whether of high-rise buildings or roads: It is the degree of interference that could be caused to the water table. Too often flooding and loss of groundwater resources occurred when underground work was conducted without adequate geological knowledge of the area.

BOSTON

Boston is more than a century and a half older than Washington, D.C. As with other coastal communities of that time, prime considerations were a safe harbor, abundant shallow groundwater, and protection from surprise attacks. The site chosen met all of these needs, and the presence of a hill in the center of the settlement provided a lookout and signal tower. Building materials were plentiful—clay for brickwork, building stone, and sand and gravel from glacial outwash for fill and for reclamation of salt marshes and tidal flats. Few other cities in North America had so varied a supply, and the early builders made good use of these materials.

Some of the practices followed over the more than 300 years of its existence show how Boston's appearance today reflects the character of local materials. For a few decades at the beginning, timber was the main resource for houses; then for all of the eighteenth century and the early years of the nineteenth, red bricks made from the abundant supplies of glacial marine clays dominated the Colonial Georgian homes. Dressed stone featured for a time in midnineteenth century, and later granite and a conglomerate added to the variety of classical stone buildings.

Tunnels for water supply, drainage, and sewage have been dug from time to time, and the lack of detailed geological information added significantly to the costs of the work as particular types of rocks and faults were encountered. One main drainage tunnel under the harbor had a section cut through Cambridge argillite where only a tenth of the length required steel supports. A further section, going through weak shales, needed steel support for almost 90 percent of its length. Some tunnels had unexpected problems with dikes that cut across prevailing strata, whereas others ran into numerous faults, some of them having fractured the surrounding rock so badly that tunneling caused water to get into the work area. Pumping was required for some months before things were brought under control.

The extensive deposits of unconsolidated material, much of it deposits from Pleistocene glaciation, and lack of knowledge of their depths led to uncertainties over foundations for multistory buildings. Trial and error was often the only approach. Reinforced concrete mats proved to be successful in clays of different textures because the building load was distributed evenly, and so there was less risk of differential ground settlement. For very heavy structures, it was found essential to follow older practices of driving displacement piles through the clays to bedrock. The Prudential Insurance Tower with 52 stories used open steel pipes driven 50 meters through the clays to the underlying argillite bedrock. Further strengthening of these steel pipes followed.

NEW YORK CITY

Underlying the city of New York are three physiographic provinces, Atlantic Coastal Plain, New England Upland, and Triassic Lowland, and between them they contain nine different foundation rock types and dozens of soils. Serious problems of foundation support in the loosely consolidated materials of the Coastal Plain are solved by the use of spread footings. In the New England Upland, a similar situation might need caissons where bedrock underlies thick beds of till. The foundation of New York City is a series of very old crystalline rocks—schist, gneiss, and marble—that cover most of Manhattan and the Bronx, and these provide excellent foundations.

All buildings, especially the skyscrapers of New York, transfer their weight to the ground beneath, so an accurate knowledge of local geological conditions is an essential safety consideration. The familiar New York skyline of very tall buildings as early as the beginning of the twentieth century, in contrast to other world cities of that time, is a direct result of the strength of the Manhattan schist foundation rock. At the same time, the rock that made skyscrapers possible added new complications when underground tunneling became necessary, and it was not long before water demands required the construction of some of the biggest tunnel systems in the United States.

Water supplies first came from shallow local wells, then from the rivers around Manhattan until cholera epidemics forced their closure. Shortage of water, however, meant that fires burned incessantly, leaving a thick pall over the downtown, a problem that soon spurred action. With commendable vision for the time, civic authorities decided to go as far as 70 kilometers upstate and build a dam on the Old Croton River. From there a series of aqueducts carried the water through ridges and over valleys in a style reminiscent of the Roman Empire and the first of its kind in the United States.

Construction began in 1837, and the water reached New York five years later, but another six years were needed to complete construction. It took 22 hours for the water to travel from the dam. This source of water was intended to meet the city's needs for centuries to come. In fact, it kept the city going for less than 50 years because population growth exceeded all expectations. The new Croton Aqueduct, three times the length of the old one, was begun in 1885, and it delivered water to the city five years later. It is still in use, but planners, knowing that it will not be adequate for very long, keep working on proposals for new sources of water.

I mentioned at the beginning of the chapter that Boston made use of geological data in planning construction as early as 1914, a rare event in urban history. New York may have been even earlier. In 1905 the City Board of Water Supply brought in a professor of geology to assist with the New Croton System, and the success of that enterprise is due in large part to that action. In other departments, too, New York made good use of geologists. Between 1920 and 1940 hydroelectric plants were built at Niagara Falls, and geologists investigated the relevant dam and reservoir sites and routes for the power lines. The interstate highway system of the 1950s and the St. Lawrence Seaway, completed in 1957, were other beneficiaries of geological advice.

By far the most important use of geologists was made between 1965 and 1975 when New York State embarked on a program of nuclear power generation. Detailed investigations were carried out on 20 possible sites as part of a federal requirement that all aspects of environmental impacts be studied. Fewer than a third of the proposed plants were ever built. Earthquake risk was one major factor in these decisions. A great deal was learned in the process about the nature of seismic activity in New York State.

Like many of the world's big cities, land reclamation soon becomes a desirable way of adding territory, especially if, like New York, the city is an island. Reclaiming marshes or parts of the ocean carries risks of problems at a later time, and New York had one close call when the Lincoln Tunnel was laid under the Hudson River but with some dependence on artificial fill at one side of the river. John F. Kennedy International Airport, formerly Idlewild Airport, is built on Jamaica Bay at the south of Queens County, an ocean bay with several islands. To date, the fill that transformed this bay into flat land has not given any problems. Terminal buildings and other airport structures are all built on piles.

There is an important property of groundwater, that is to say, the water that is stored underground either in hollowed sections of impervious rock or in the interstices between rocks. It is its temperature, which is usually the same as the average air temperature at the same location. In hot summers this cooler water can be pumped into the air-conditioning systems of buildings to reduce costs, then returned to ground. Some theaters in New York take advantage of this cost-saving arrangement, and to avoid the risk of contamination, they circulate the water in its own closed system.

The variety of construction activities carried on in New York, especially in the early decades of the twentieth century, provided a better picture of the city's geology than any other method of studying it. Housing, water and sewer systems, the many bridges for this island city, subway tunnels, and so on, all demanded precise understandings of the rocks beneath before work could begin. Much of it was learning on the job, but fortunately, as was mentioned in the case of the water aqueducts, city authorities had the good sense to employ geologists as consultants. All the large structures of the city have foundations excavated down to bedrock. Foundations for the Twin Towers of the World Trade Center, which is a good example of a skyscraper, go down into bedrock to a depth of eight stories.

The Broadway Subway System was built at the beginning of the twentieth century, and it required tunneling through the Manhattan schist. The engineers decided that the rock was so strong and so attractive that they stockpiled it. Years later it was used to construct the Gothic buildings of New York's City College. A similar thing happened when marble, or rather sand residue from decomposition of the marble, was found in excavations for Interstate 95. This material was also stockpiled and later found to be exactly the right kind of base course material for pavements because it has the property of preventing the buildup of moisture under the pavement, a condition that would give rise to frost heave in cold weather.

In summary, the different kinds of rocks under the city date from the earliest geological eras right up to the Holocene, and they have different strengths and suitabilities. City planners redraft their construction codes as new discoveries about rocks are made. Over the years the shorelines have changed as sea areas become land after reclamation, and land surfaces with their watercourses are obliterated under road surfaces and buildings. Full details of the original sites are an essential part of planning, whatever structures might be in place at a given time. Armed with this information, together with the history of all the changes that have taken place, good and safe plans can be made for new structures.

New York City has rocks of all kinds in its buildings, many of them imported from other countries by the architects who designed the structures. The study of these building stones reveals how well or how badly they responded to the erosive forces of rain and wind and thus provides information for future uses. Their types and sizes change from decade to decade, so they also give us a historical record of architectural styles. By looking at buildings in this way, we get a lot of firsthand information about the geology of places we would never normally see. The names that are popularly used for building stones are the ones I will use here, even though they may not be correct geologically, just as flower names frequently differ from the proper botanical terms.

Grand Central at 42nd and Park Avenue, in downtown New York, is one of the great railway stations of the world, with a history going back to the earliest days of the city. Indiana limestone from western Indiana was the stone selected for the exterior of the building. Several producers of building stones were asked to submit samples so they could be exposed for some time to see how well they stood up to New York's weather. Indiana limestone was then chosen for its durability and lower cost. Over the years both here and in the hundreds of other locations throughout the city where it was used, this rock suffers from the effects of acid rain, which easily eats into the limestone, leaving a rough, granular surface. Stony Creek granite from Connecticut was selected for the elevated highway and Italian marble for the interior floors. The marble did not stand up well and was later replaced with other material.

St. Patrick's Cathedral is the largest Catholic cathedral in North America. Its design is Gothic, and throughout the building, the stone of choice is local marble, quarried from different places in New York City and its environs. One stone used for the main facade is called "snowflake marble" because of its white color. Construction began in 1858. Over the years, major problems of stone disintegration appeared, and one substitute rock that was used extensively was Georgia marble from Tate in Georgia. Other rock types used here were marble from Massachusetts for the outer houses of the church and granite from Maine for the base of the main building. The New York Public Library is another building of special interest. In all the main parts of the building, marble rock from one source, Vermont, was used. It, too, suffered deterioration and change in color in the course of the twentieth century, and several attempts at restoration are evident if one examines it closely.

WASHINGTON, D.C.

The nation's capital is the first and largest planned city in the country. Its location along the Fall Line boundary between the Atlantic Coastal Plain and the Piedmont Plateau, at the head of navigation on the Potomac River, ensured ocean access (Figure 10.3). Since the Fall Line was also the easiest place for bridging rivers, highways were built to the north and south, linking the city with inland areas. A saucer-shaped rim of hills made the selected site good for defense, the main consideration for the young nation of the 1790s. But unknown to President George Washington, who selected the final site, his choice guaranteed that the city of Washington would have a very complex geological underpinning at the junction of two vastly different rock systems.

The western part of the city and most of the western and northern suburbs are on the Piedmont Plateau, an upland area underlaid by complex, ancient metasedimentary and metaigneous rocks, 400 million or more years old. These crystalline rocks are covered by soil and weathered rock to depths as much as 50 meters, making them difficult to analyze and hard to excavate. Their strengths are highly variable because of the mineral content of the parent rocks as well as the degree of weathering to which they were exposed. These weathered materials range from soillike to rocklike within quite short lateral and vertical distances. This is important information for the building industry. Washington has many high office towers, and they need dependable foundations.

Toward the southeastern sections of the urban area lie the Atlantic Coastal

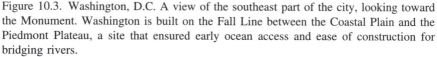

Figure 10.3. Washington, D.C. A view of the southeast part of the city, looking toward the Monument. Washington is built on the Fall Line between the Coastal Plain and the Piedmont Plateau, a site that ensured early ocean access and ease of construction for bridging rivers.

Plain's fluvial and marine strata of comparatively young age, in depths as much as 450 meters. Water is abundant, and shallow-dug wells met the needs of residents prior to the Civil War. In 1963 the aqueduct from Great Falls provided all the water that was needed. The Potomac River still provides the city's water supplies. During the early years of the city, local building materials played a large part in the design of architectural styles and types of buildings. Stone was taken from the crystalline rocks common to the Piedmont terrain and sedimentary rocks from the Coastal Plain. Brick manufacturing used clay from Coastal Plain sediments. Development was restricted in lowland areas along the Potomac and Anacostia Rivers because of the danger of flooding.

The history of the Washington Monument is a good example of the problems associated with building on any of the low-lying areas of the city. When first built in 1848, its foundation was set in sand and clay 10 meters below ground level and carried to a height of 47 meters. Construction stopped in 1854 due to lack of private contributions and was not resumed until 1880. During this time the Monument settled and tilted northward due to some yielding on the part of a hitherto unrecognized clay bed 10 meters below the foundation. Underpinning had to be inserted to increase the foundation's area by 50 percent and, at the same time, bring the axis back into plumb. The final structure of 169 meters, which was finished in 1884, settled a further 15 centimeters over the succeeding

50 years, but because of the additional concrete underpinning, it remained vertical.

In modern times this great variety of rocks was more of a problem than an asset. In the 1970s, construction of tunnels for the underground transit system brought city engineers face to face with the kinds of rocks that lay at deeper levels. There was little experience with Piedmont rocks before this. For economic reasons, both trains and stations had to be underground, and wherever possible, tunnel-boring machines worked in Potomac rocks of the Coastal Plain. Because of the geological history of this part of the nation, a great deal of faulting occurred at different times. Thus it happened that on more than one occasion, the tunnel-boring machine encountered a mixture of bedrock and sediments. Progress was slow and expensive in these settings.

As the nation's capital, Washington has numerous stately buildings, many of them constructed with stones from different states. We do not usually pay much attention to this aspect of their construction because we are more interested in their architecture and function. The stones, however, have quite a story to tell if we take the time to look closely: Their ability over time to cope with physical and chemical erosion from rain and wind reveals a lot about the nature of the rocks being used. They may also provide useful information on what to expect in the places across the country from which they came as rocks of the same type are subjected to erosion.

Constitution Hall was built in 1929 using limestone rock from Alabama. This rock, like all carbonates, is sensitive to water erosion, especially if the water has any chemical elements that might speed up the erosive process. This particular stone has stood up well over the past 70 or more years; but salting of the steps and landing over a number of winters took its toll, and substantial restoration was needed in the 1980s, including replacement of some of the limestone. Where overhangs protect a wall from rain, there is no evident damage to the limestone rock beneath. The Renwick Museum, built in 1859, used sandstone rock from New Jersey, which, like limestone, is sensitive to acidic water. In this case the sandstone had to be replaced a hundred years later after earlier attempts at restoration failed.

Memorial Continental Hall, the original building of the Daughters of the American Revolution, was built in 1909 using fine-grained marble from Vermont. Over the years since then it was sandblasted and hydraulically cleaned at different times, but by the 1980s, the rock was in such a state of deterioration that much of the surface had to be replaced with coarse-grained marble from Georgia. Acid rain and hail are the main erosive forces causing the damage. As was the case with other buildings, sheltered areas under overhangs or inside a porch remained in good condition. Red Cross buildings, more than 20 years later, used the same type of rock as Memorial Hall and had the same results— but with an addition: Leaching from overhead lamps attacked some of the carbonate in the marble, and extensive cracking ensued.

Jefferson Memorial, often considered the most beautiful of all U.S. monuments perhaps because of its location near the Tidal Basin, was built in 1943 using fine marble from Vermont for the exterior and coarse-grained Georgian marble for the interior. It is a pity that the places where the two kinds of marble were used had not been reversed because in that case much less pollution damage would be visible. Coarse-grained marble has a much greater resistance to erosion. A surprising amount of damage can be seen today even though the building is so young. The polish is gone from many of the exterior vertical surfaces, with, once again, much less destruction on protected surfaces.

REFERENCES FOR FURTHER STUDY

Barlow, Elizabeth. *The Forests and Wetlands of New York City*. Boston: Little, Brown, 1971.

Froelich, A. J., and E. G. Otton. *Geologic and Hydrologic Reports for Land-Use Planning in the Baltimore-Washington Urban Area*. Washington, D.C.: U.S. Geological Survey Circular 806, 1980.

Hanshaw, Penelope M., ed. *Environmental, Engineering, and Urban Geology in the United States*. Vols. 1–2. Washington, D.C.: American Geophysical Union, 1989.

Johnston, P. M. *Geology and Ground Water Resources of Washington, D.C. and Vicinity*. Washington, D.C.: U.S. Geological Survey Water Supply Paper 1776, 1964.

Legget, Robert F. *Cities and Geology*. New York: McGraw-Hill, 1973.

Legget, Robert F., ed. *Geology under Cities*. Boulder, CO: Geological Society of America, 1982.

Schuberth, C. J. *The Geology of New York City and Environs*. Garden City, NY: Natural History Press, 1968.

Weidner, C. H. *Water for a City*. New Brunswick, NJ: Rutgers University Press, 1974.

GEOLOGICAL TIME CHART

Era	Period	Epoch	Time in Millions of Years before Present
CENOZOIC	Quaternary	Holocene	0.01
		Pleistocene	1.5
	Tertiary	Pliocene	5
		Miocene	24
		Oligocene	37
		Eocene	58
		Paleocene	65
MESOZOIC	Cretaceous		144
	Jurassic		208
	Triassic		245
PALEOZOIC	Permian		286
	Pennsylvanian		320
	Mississippian		360
	Devonian		408
	Silurian		438
	Ordovician		505
	Cambrian		570
PRECAMBRIAN			

U.S. CENSUS DIVISIONS AND REGIONS

Census Division	Region	States
WEST	Pacific	Alaska
		California
		Hawaii
		Oregon
		Washington
	Mountain	Arizona
		Colorado
		Idaho
		Montana
		Nevada
		New Mexico
		Utah
		Wyoming
MIDWEST	West North Central	Iowa
		Kansas
		Minnesota
		Missouri

Census Division	Region	States
		Nebraska
		North Dakota
		South Dakota
	East North Central	Illinois
		Indiana
		Michigan
		Ohio
		Wisconsin
NORTHEAST	Middle Atlantic	New Jersey
		New York
		Pennsylvania
	New England	Connecticut
		Massachusetts
		Maine
		New Hampshire
		Rhode Island
		Vermont
SOUTH	South Atlantic	Delaware
		Florida
		Georgia
		Maryland
		North Carolina
		South Carolina
		Virginia
		Washington, D.C.
		West Virginia
	East South Central	Alabama
		Kentucky
		Mississippi
		Tennessee

Census Division	Region	States
	West South Central	Arkansas
		Louisiana
		Oklahoma
		Texas

APPENDIX C _____

ACTIVE AND POTENTIALLY ACTIVE VOLCANOES IN THE UNITED STATES

Alaska and Aleutian Islands

Wrangell

Mount Spurr Redoubt

Iliamna

Mount Mageik

Novarupta

Mount Ugashik

Aniakchak

Mount Emmons

Kiska

Mount Cerberus

Tanaga

Great Sitkin

Korovin

Amukta

Carlisle

Kagamil

Okmok

Makushin

Alaska (*cont.*)	Mount Hayes
	Augustine
	Mount Martin
	Trident
	Yantarni
	Mount Veniaminof
	Mount Dutton
	Little Sitkin
	Mount Gareloi
	Kanaga
	Kasatochi Island
	Pyre Peak
	Yunaska
	Mount Cleveland
	Mount Vsevidof
	Bogoslof
	Akutan Peak
	Westdahl Peak
	Shishaldin
	Fisher Dome
	Isanotski Peaks
Arizona	San Francisco Field
California	Medicine Lake
	Lassen Peak
	Long Valley
	Mount Shasta
	Clear Lake
	Coso Peak
Hawaii	Loihi
	Mauna Loa
	Mauna Kea
	Haleakala
	Kilauea

Hawaii (*cont.*)	Hualalai
	Kohala
Idaho, Montana, Wyoming	Craters of the Moon (Idaho)
	Yellowstone (Idaho, Montana, Wyoming)
New Mexico	Bandera Field
Oregon	Mount Hood
	Mount Jefferson
	Three Sisters
	Crater Lake
	Newberry
Washington	Mount Baker
	Glacier Peak
	Mount Rainier
	Mount St. Helens
	Mount Adams

_____

METRIC AND IMPERIAL MEASURES

Length

1 inch	equals 2.54 centimeters
1 centimeter	equals 0.39 inches
1 foot	equals 0.3 meters
1 meter	equals 3.28 feet
1 mile	equals 1.61 kilometers
1 kilometer	equals 0.62 miles

Area

1 square inch	equals 6.45 square centimeters
1 square centimeter	equals 0.15 square inches
1 square foot	equals 0.09 square meters
1 square meter	equals 10.76 square feet
1 square mile	equals 2.59 square kilometers
1 square kilometer	equals 0.39 square miles

Volume

1 cubic inch	equals 16.38 cubic centimeters
1 cubic centimeter	equals 0.06 cubic inches
1 cubic foot	equals 0.028 cubic meters
1 cubic meter	equals 35.3 cubic feet

| 1 cubic mile | equals 4.17 cubic kilometers |
| 1 cubic kilometer | equals 0.24 cubic miles |

Temperature

- To change Fahrenheit to Celsius, subtract 32 degrees and divide remainder by 1.8.
- To change Celsius to Fahrenheit, multiply by 1.8 and add 32 degrees.

GLOSSARY

accretion. Addition of new crust to a continent, often through action by ocean crust ferrying rock from islands.

acid rain. Rainwater that is so acidic that it damages the environment; caused by pollution from industry and automobiles.

adiabatic. Change of sensible temperature within a gas because of compression or expansion without gain or loss of heat from outside.

aggregate. A mixture of gravel and crushed stone.

air mass. A large area of air that has nearly uniform characteristics such as water vapor concentration and temperature at any given altitude.

alluvial sediments. Sediments deposited by a stream or flowing water.

anthracite. Grade of coal that is very high in fixed carbon content with little volatile matter; also called hard coal.

aquifer. An underground body of saturated rock or sediment that is both porous and permeable enough to provide usable quantities of water.

area studies. Studies of particular countries or groups of countries.

arete. Narrow, steep-sided ridge between two cirques.

argon. Inert gas that forms about 1 percent of the earth's atmosphere. Its presence in magma is used to date the rock after it cools.

artificial reef. Obstruction placed offshore to make waves break earlier than they normally would.

asbestos. A fibrous incombustible mineral traditionally used as a heat-resistant or insulating material.

ash flow. A concentrated dispersion of hot volcanic fragments such as pumice; also known as pyroclastic flow.

asteroids. Small planetary objects orbiting the sun. Most of them are found between the orbits of Mars and Jupiter.

asthenosphere. Soft layer of the upper mantle; it lies under the lithosphere.

atoll. A ring-shaped coral reef surrounding a central lagoon.

avalanche. A mass of ice or snow moving rapidly down a mountain.

avulsion. The process by which a stream changes course by a sudden diversion.

backswamp. A waterlogged area adjacent to a river containing fine-grained sediments.

badlands. Landscape of deep gullies and ravines produced by extensive erosion.

barrier island. A long, narrow island, mainly sand, built by waves near a beach and separated from it by a lagoon.

basalt. A dark-colored, dense igneous rock formed from molten rock. Ocean floor (oceanic crust) is made of basalt.

basal water. A body of freshwater that floats on saltwater within an aquifer.

basement rock. Deeply eroded metamorphic bedrock, usually covered by younger sedimentary rocks.

batholith. Large body of intrusive igneous rock, usually granite in composition.

bayou. Small stream found in an abandoned river channel or in a delta.

bedrock. The solid rock that lies under the soil.

bituminous. Grade of coal that has substantial content of volatile material; also called soft coal.

blowdown. A very strong wind capable of uprooting trees.

bluff. A high, steep bank or cliff.

borehole. A hole drilled for exploration.

breakwater. A structure to break the force of waves.

butte. Steep-sided, flat-topped hill or mountain.

calcite. The most common carbonate mineral and the main component of limestone.

caldera. A circular depression at the top of a volcano, with a diameter many times larger than the volcanic vent.

canyon. A steep-sided bedrock valley with a narrow floor.

caprock. A hard layer of rock overlying softer strata.

carbonate rocks. Minerals like calcite or dolomite that are compounds of calcium or magnesium.

carbon dioxide. A gas that occurs naturally in the atmosphere and is formed by respiration.

catchment. The total area of a drainage basin.

chloride. Any compound of salt with another element or group of elements.

chlorofluorocarbon (CFC). Gaseous compound found in older refrigerators and considered to be harmful to the earth's upper atmosphere ozone layer.

chromium. Hard white metallic element used in alloys as a decorative electroplated coating.

cirque. Horseshoe-shaped, steep-walled valley at the head of a glacier.

clay. Particles of sediment smaller that 0.004 millimeters in diameter.

climate. A generalized statement about weather conditions at a given place based on statistics taken over a long period of time.

cloud seeding. Use of ice crystals in storm clouds with a view to reducing temperature.

coal. A rock consisting of carbon compounds formed over time from plant remains.

coastal plain. A gently sloping plain at the edge of a continent, often former continental shelf land that emerged from the sea.

cobalt. A silvery-white magnetic metallic mineral occurring naturally as a mineral.

continental drift. The theory advanced by Alfred Wegener and others early in the 1900s to explain the movement of continents over geological time.

continental shelf. Gently sloping land on the sea floor adjacent to a continent, ending at its outermost edge at the continental slope.

convection currents. Circular motion within a fluid created when warmer materials rise and cooler ones sink.

copper. A malleable metallic element often used as an electrical conductor.

coral reef. Rocklike accumulation of carbonate material secreted by corals.

coulee. A deep ravine with steep sides.

crater. A bowl-shaped depression at the top or flank of a volcano.

craton. The stable interior of a continental plate.

crust. The thin, solid shell of rock that forms the outermost layer of the earth.

crystalline rocks. Large individual mineral grains in igneous or metamorphic rocks.

cumulonimbus cloud. Exceptionally dense and vertically developed cloud, either isolated or as a wall of clouds.

cyanide. Any of the poisonous salts from hydrocyanic acid.

cyclone. Center of low pressure with oval-shaped ground plan and inward-moving air currents.

debris flow. Rapid mass downslope movement of muddy water and sediment of a wide range of sizes, including boulders.

delta. The triangular-shaped mass of sediments at the mouth of a river.

deposition. The process of depositing sediments.

dike. A near-vertical minor igneous intrusion.

dip. Angle between an inclined rock surface and a horizontal reference plane.

dolomite. Sedimentary rock composed of calcium magnesium carbonate.

drumlin. A hill of glacial till, oval in shape and smoothly rounded on top.

dune. Mound of unconsolidated granular material of sand size and durable composition.

dust storm. Clouds of dust in winds from a turbulent air mass.

earthquake. Trembling or shaking of the surface of the ground due to seismic activity.

earth science. The scientific study of the earth, its materials, biography, and environment in space.

earth slump. Collapse of part of a mountain slope or bluff.

ecology. Organisms in their relations to one another and their environment.

endangered species. A species that has been reduced in number to the point where its total extinction is a possibility.

environment. The physical environment and general conditions within which people live and work.

environmentalist. One who is actively concerned with preservation and improvement of the environment.

epicenter. A point on the surface of the earth directly above the underground source of an earthquake.

epoch. A third-order geological time unit.

era. A first-order geological time unit, that is to say, covering the largest span of time.

erosion. The wearing away of the land, mainly by rain and running water.

erratic. A stone deposited by a glacier far from its point of origin.

esker. An elongated ridge of glacial till.

evapotranspiration. The combined loss of water to the atmosphere from evaporation and transpiration.

fall line. The boundary between resistant rocks of older land and the weaker strata of plains.

fault. Sharp break in bedrock with displacement of adjacent rock.

feldspar. Any one of a group of alumina silicates, the most abundant material in the earth's crust.

ferrous scrap. Used metals with high iron content.

finger lake. Long, narrow lake in a glacial trough.

firestorm. An intense conflagration into which surrounding air is drawn with great force.

flash flooding. A sudden local flood of short duration and great volume, usually caused by torrential rain.

flood plain. Area of low, flat ground on one or both sides of a stream; usually underlaid by alluvium from past flooding.

fossil. The remains of a prehistoric plant or animal.

fossil fuels. A collective term for coal, petroleum, and natural gas, all resources that are used for providing energy.

fumarole. A vent from which volcanic gases escape.

gene pool. The sum of all genes in an interbreeding population of plants or animals.

geological fault. Sharp break in bedrock with displacement of one side relative to the other.

geology. Study of the solid earth including its origin, history, and composition.

geomorphology. Study of the surface of the earth, including the history of its landforms and processes at work today.

geophysicist. A geophysics specialist, that is to say, one who specializes in the physics of the earth and its environment.

geothermal. Relating to the flow of heat from the interior of the earth to the surface.

geyser. A vent from which hot water and steam erupt in a volcanic area.

glacial drift. A general term for all the different kinds of sediments and debris deposited in connection with the last ice age.

glacial lake. A lake that was created by the action of ice during the last ice age.

glacial outwash. Layers of sand and gravel deposited by glacial meltwater streams.

glacier. Large mass of ice, air, water, and rock debris formed on land, moving slowly downhill.

Global Positioning System. A worldwide system of satellites and ground stations that make possible precise location of places on the ground.

global warming. Rise in average temperatures due to increased consumption of fossil fuels and consequent increase of carbon dioxide in the atmosphere.

gneiss. A metamorphic rock rich in quartz and feldspar.

gold. A yellow metallic element that is resistant to chemical reaction, occurring naturally in quartz veins.

granite. An intrusive igneous rock that contains large amounts of quartz and smaller quantities of several other minerals.

gravel. Unconsolidated deposits of rounded stones or boulders.

greenhouse gases. Gases, especially carbon dioxide, that contribute to the greenhouse effect, that is to say, helping to trap heat within the atmosphere.

groundwater. Water held in soil or rock underground.

gully. A deep V-shaped valley formed by a young stream.

guyot. A flat-topped submarine peak; originally a volcano above sea level.

habitat. The environment or place where a plant or animal normally lives and grows.

hanging valley. A stream that was truncated by the action of a valley glacier.

Holocene. The youngest epoch of the Quaternary, about 10,000 years ago.

horn. A pyramidal peak left after erosion on all sides of a mountain.

hornblende. A dark or green mineral composed of silicates and occurring in igneous or metamorphic rocks.

hot spot. A center of continuing volcanic activity within a lithospheric plate.

hot spring. Jets of boiling water erupting from an underground source of lava.

hurricane. A tropical cyclone; most hurricanes occur along the eastern coasts of the United States and in the Caribbean.

hydraulic head. The difference in water level between two points in the water table.

hydrogeology. The study of the effect of geology on water at the surface of the earth and underground.

ice age. The span of geological time, usually 1 or 2 million years, during which large parts of North America and other continents were covered with ice.

ice sheet. Large, thick area of ice moving out in all directions from a center of accumulation.

igneous rock. Rock solidified from a molten state, usually through eruption of magma.

impermeable. Not allowing water or gas to pass through it.

interstices. Tiny spaces within rocks or soil.

isostasy. The response of the earth to the imposition or removal of large weights.

jet stream. High-elevation air flow moving at great speed within a tubelike form.

karst topography. A landscape of caves, disappearing streams, and sinkholes, all the result of groundwater dissolving limestone.

lacustrine deposits. Sediments deposited in lakes.

lagoon. A large body of saltwater near the ocean.

landfill. A place where solid waste generated by humans is buried.

landslide. Rapid movement of masses of bedrock or soil on steep mountain slopes.

laser. A device that generates an intense beam of single wavelength light.

lava. Magma emerging from the earth and exposed to air or water.

lead. Heavy, ductile, metallic element occurring naturally in other minerals.

levee. An embankment bordering a stream or river to prevent flooding when water level is high.

lignite. Low grade of coal with a heat value between coal and peat.

limestone. Sedimentary rock in which calcite is the dominant mineral.

liquefaction. Transformation of soil or unconsolidated sediments into a liquid state, usually as a result of an earthquake.

lithosphere. The strong, brittle, outermost part of the earth. In the case of the oceans it lies immediately below and supports the oceanic crust.

loess. Fine-grained sediment deposited after being transported by wind.

longshore drift. Shore-parallel movement of sediment in one direction.

magma. Molten rock about to emerge from the earth.

magma chamber. A subsurface pool of magma.

magnesium. A silvery metallic element occurring naturally in dolomite and an essential element in living organisms.

magnetite. An important ore of iron.

magnetometer. An instrument for measuring magnetic strength, especially that of the earth.

mantle. A thick layer of dense rock below the crust of the earth and above the core.

marble. A metamorphic rock derived from limestone or dolomite.

Mercalli Scale. An intensity scale based on the effects of earthquakes developed by G. Mercalli in 1902, using 12 intensity levels. Later refined as the Modified Mercalli Scale to fit building and social conditions in the United States.

mercury. A toxic silvery-white metallic element used in barometers and other instruments.

mesa. Small table-topped plateau.

metamorphic rocks. Rocks that have been altered either physically or chemically as a result of very high heat or pressure.

mica. Any one of a group of silicate minerals with a layered structure.

microburst. A severe thunderstorm in the form of a downdraft several kilometers wide.

midocean ridge. An underwater mountain chain that forms at the divergent margin in the middle of a widening ocean.

mineral. A natural inorganic solid substance with a definite chemical composition.

Miocene Epoch. A third-order geological time unit.

molybdenum. A silver-white metallic element used in steel to give strength and provide resistance to erosion.

moraine. A pile of unsorted glacial drift.

mud pot. A hot spring that has been mixed with mud.

natural gas. A flammable gas found in the earth's crust.

nickel. A natural malleable metallic element used in steel production.

nonrenewable resources. Those with fixed quantities in the earth's crust such as coal and petroleum.

nuclear waste. Radioactive waste material from nuclear-powered electrical power stations.

oceanic crust. Relatively thin, compared with continental, dense layer of basalt covering ocean basins.

oil. A thick liquid found in the earth's crust and refined into gasoline and other products.

old-growth forest. A mature forest that has never been felled.

opencast mining. Mining by excavation from the surface directly above.

orbit. Path followed by a planet as it revolves around the sun.

orogen. The mass of rocks of different kinds pushed upward by subduction action at continental edges.

orographic lifting. The action of mountains in forcing air to rise.

outcrop. Rock that appears at the surface of the earth.

outwash plain. Flat, gently sloping plain formed by meltwaters at the front edge of an ice sheet.

overburden. Surface rock and soil that lies at the surface above a mineral deposit.

ozone layer. A layer in the stratosphere in which a concentration of ozone is created by the sun's rays.

Paleozoic era. A first-order geological time unit.

Pangea. Supercontinent that formed as a result of many volcanic activities and broke apart in a worldwide rifting over 200 million years ago.

period. A second-order geological time unit.

permafrost. Ground that remains frozen continuously for several years.

permeable. Allowing surface and groundwater to pass easily.

petrified wood. Ancient wood remains that have been changed into a stony substance.

petroleum. Oil or natural gas.

phosphorus. A nonmetallic element occurring naturally in various phosphate rocks.

piedmont. Land at the base of a mountain range.

pingo. A conical mound with a core of ice, found in permafrost areas of Alaska.

placer deposits. Deposits of sand or gravel in a stream containing valuable minerals such as gold.

plagioclase. A series of feldspar minerals that form glassy crystals.

plateau. An upland surface, flat and mainly horizontal.

plate tectonics. Global model involving a small number of semirigid plates that float on an underlying mantle.

platinum. A ductile, malleable metallic element used in making jewelry and laboratory apparatuses.

plutonic rock. Igneous rock that formed when magma cooled and hardened beneath the earth's surface.

potassium. A soft metallic element occurring naturally in seawater. It is essential for living organisms.

Precambrian era. A first-order geological time unit.

proglacial lake. A lake in front of a glacier or ice sheet.

pyroclastic flow. *See* **ash flow**.

quartzite. A metamorphic rock formed by metamorphism of sandstone.

Quaternary period. A second-order geological time unit.

radioactive decay. The process by which an unstable isotope changes into another isotope by emitting particles of energy.

radon. The only radioactive gaseous element; presence often indicative of buried uranium.

remanent magnetism. The permanent magnetization of rocks at the time they cooled and solidified from magma.

renewable resources. Those resources that can be renewed, such as forests.

Richter Scale. A measure of the quantity of energy released by an earthquake. The scale is logarithmic; that is, each number represents 10 times the energy of the previous number.

ring of fire. Continental margins around the Pacific Ocean because of the frequency of earthquakes and volcanic eruptions there.

salt plain. An area of land saturated with salt and evident by its whitish color.

sand dune. A hill or ridge of well-sorted sand that has been shaped by wind erosion.

sandstone. A sedimentary rock with mineral particles the size of sand.

schist. A leaflike metamorphic rock.

seamount. A submarine peak that does not rise above sea level.

sedimentary rock. Rock formed from the accumulation of sediments over a long period of time.

seismic. Relating to an earthquake or other vibrations of the earth and its crust.

shale. A fine-grained sedimentary rock.

shield volcano. A broad-domed volcano with sloping sides.

silicon. A nonmetallic element found in abundance in the earth's crust and used in electronic devices.

silt. Sediment particles measuring less than 0.06 millimeters in diameter.

silver. A malleable precious metallic element used mainly in coins.

sinkhole. A surface depression as a result of underlying limestone rock being dissolved by groundwater.

slate. A fine-grained rock metamorphosed from sedimentary rock.

smog. Air pollution in the lower atmosphere over urban areas.

stratovolcano. A steep-sided conical volcanic peak.

strike. Compass direction of a line of intersection between an inclined rock plane and some reference horizontal plane.

strip mining. A form of opencast mining in which strips of land are mined and later filled in as the next strip is opened up.

subduction. Downward movement of a lithospheric plate at its edge in order to pass beneath the edge of the adjoining plate.

sulfur. A pale-yellow nonmetallic element used in making matches and sulfuric acid.

surficial deposits. Loose sediments lying at or near the surface of the earth.

swelling soils. Soils that expand when they are wet and contract when dry.

tar pit. Pool of bitumen, frequently containing the remains of animals.

tectonic plates. *See* **plate tectonics**.

terrane. A crustal rock unit with distinctive lithologic properties that reflect its geological history.

thermokarst topography. Topographic depressions in permafrost landscapes.

tholeiitic. Basaltic lava with high concentrations of silica; the most abundant of the basaltic group.

till. An unsorted mixture of clay, sand, and gravel deposited by a glacier.

tin. A metallic malleable element used mainly in alloys and in the plating of iron and steel.

topography. The shape and height of the earth's surface.

tornado. Small storm that forms beneath a cumulonimbus cloud and carries extremely high winds.

trench. Long, deep depression in the ocean floor marking the place where oceanic lithosphere subducts beneath continental lithosphere.

tropical cyclone. Intense traveling cyclone with high winds and heavy rainfall.

tsunami. A sea wave caused by a volcanic eruption; also known as a seismic sea wave.

tundra. Treeless Arctic regions frequently with underlying permafrost.

tungsten. A dense metallic element with a high melting point, used mainly for the filaments of electric lights.

unconsolidated sediments. Loose or uncemented sediments.

uniformitarianism. The theory that present rates of change have always been the rates of change in the past.

uranium. A heavy radioactive metallic element capable of nuclear fission.

U-shaped valley. A valley with a broad floor and steep sides formed by glacial erosion.

volcanic rock. Rock that is formed by volcanism.

volcanism. General term for volcanic activity that erupts onto the surface; also known as vulcanism.

water table. Upper surface of the saturated zone of soil or rock.

wave-cut scarp. A steep-sided hill or bluff formed by the erosive action of ocean waves.

yazoo stream. A stream that enters the floodplain of a larger river and is forced by the levees to run parallel with the main river.

zinc. A white metallic element used in batteries and in galvanizing sheet iron.

BIBLIOGRAPHY

THE GEOLOGY OF NORTH AMERICA

Part of the Decade of North American Geology Project, Published by the Geological Society of America, Inc., Boulder, Colorado, 1988–1991

Volume A, *The Geology of North America: An Overview*, A. W. Bally and A. R. Palmer, eds.

Volume C-2, *Precambrian Conterminous U.S.*, J. C. Reed, Jr., et al., eds.

Volume D-2, *Sedimentary Cover: North American Craton, U.S.*, L. L. Sloss, ed.

Volume F-2, *Appalachian-Ouachita Orogen in the United States*, R. D. Hatcher, Jr., G. W. Viele, and W. A. Thomas, eds.

Volume G-1, *The Cordilleran Orogen: Alaska*, G. Plafker, D. L. Jones, and H. C. Berg, eds.

Volume G-3, *Cordilleran Orogen: U.S.*, B. C. Burchfiel, P. W. Lipman, and M. L. Zoback, eds.

Volume I-2, *The Atlantic Continental Margin: U.S.*, R. E. Sheridan, and J. A. Grow, eds.

Volume K-2, *Quaternary Non-glacial Geology: Conterminous U.S.*, R. B. Morrison, ed.

Volume K-3, *North America and Adjacent Oceans during the Last Deglaciation*, W. F. Ruddiman and H. E. Wright, Jr., eds.

Volume L, *The Arctic Ocean Region*, A. Grantz, J. F. Sweeney, and G. L. Johnson, eds.

Volume M, *The Western North Atlantic Region*, P. R. Vogt and B. E. Tucholke, eds.

Volume N, *The Eastern Pacific Region*, E. L. Winterer, D. M. Hussong, and R. W. Decker, eds.

Volume O-1, *Surface Water Hydrology*, M. G. Wolman and H. C. Riggs, eds.

Volume O-2, *Hydrogeology*, W. R. Back, J. S. Rosenshein, and P. R. Seaber, eds.

Volume P-2, *Economic Geology: U.S.*, R. B. Taylor, D. D. Rice, and H. J. Gluskoter, eds.

CREDITS FOR ILLUSTRATIONS

I.1   American Institute of Professional Geologists, Arvada, Colorado.

I.2   Robert H. Fickies, Geological Survey, State of New York.

I.3   *Facing Hazards*, USGS Circular, 1990, p. B43.

I.4   Kenneth S. Johnson, Oklahoma Geological Survey.

I.5   Robert H. Fickies, Geological Survey, State of New York.

I.6   *Look Before You Build*, USGS Circular, 1975, p. 14.

I.7   *Look Before You Build*, USGS Circular, 1975, p. 17.

1.1   *Geology of New York*, New York State Geological Survey, map, 1991.

1.2   *Geology of New York*, New York State Geological Survey, map, 1991.

2.1   Geological Society of America, *Geology of North America*, Vol. A.

2.2   *Geology of New York*, New York State Geological Survey, 1991, p. 14.

2.3   *Geology of New York*, New York State Geological Survey, 1991, p. 14.

2.4   *Geology of New York*, New York State Geological Survey, 1991, p. 13.

2.5   *Geology of New York*, New York State Geological Survey, 1991, p. 15.

2.6   *Facing Hazards*, USGS Circular, 1990, p. B5.

2.7   *Facing Hazards*, USGS Circular, 1990, p. B9.

2.8   American Institute of Professional Geologists, Arvada, Colorado.

2.9   American Institute of Professional Geologists, Arvada, Colorado.

3.1   Geological Society of America, *Geology of North America*, Vol. N, GSA, p. 384.

3.2   National Weather Service, Alaska.

3.3   *Look Before You Build*, USGS Circular, 1975, p. 12.

3.4   American Institute of Professional Geologists, Arvada, Colorado.

3.5   Geological Survey of Canada.

3.6   Geological Survey of Canada.

4.1   J. D. Griggs, USGS.

4.2   Geological Society of America, *Geology of North America*, Vol. N, GSA, p. 189.

4.3   J. D. Griggs, USGS.

4.4   *USGS Professional Paper*, 1998, p. 935A.

4.5   U.S. Navy.

4.6   Ed. Wolfe, USGS.

4.7   J. D. Griggs, USGS.

4.8   *USGS Yearbook*, 1998.

4.9   Geological Society of America, *Geology of North America*, Vol. N, GSA, p. 258.

5.1   *Rocks and Minerals*, July/August 1991, p. 263.

5.2   Portland, Oregon, Convention and Visitors Bureau.

5.3   Portland, Oregon, Convention and Visitors Bureau.

5.4   Patrick T. Pringle. Courtesy of *Rocks and Minerals* Magazine (July/August 1991): 268.

5.5   David K. Norman. Courtesy of Washington State Department of Natural Resources. *Washington Geology* (September 1998): 14.

5.6   Hugh Shipman. Courtesy of Washington State Department of Natural Resources. *Washington Geology* (March 1997): 21.

5.7   *Washington Geology*, March 1997, p. 17.

5.8   Portland, Oregon, Convention and Visitors Bureau.

5.9   *USGS Professional Paper*, 1997, p. C38.

5.10   *USGS Professional Paper*, 1997, p. C40.

5.11   *USGS Professional Paper*, 1997, p. C41.

5.12   American Institute of Professional Geologists, Arvada, Colorado.

5.13   San Francisco, Convention and Visitors Bureau.

5.14   American Institute of Professional Geologists, Arvada, Colorado.

5.15   Mark Milligan, Utah Geological Survey.

6.1   Courtesy of Connecticut Department of Environmental Protection.

6.2   Geological Survey of Canada.

6.3   Robert H. Fickies, Geological Survey, State of New York.

6.4   Geological Survey of Canada.

6.5   Robert H. Fickies, Geological Survey, State of New York.

6.6   Michael Bell, *The Face of Connecticut* (Hartford, 1988), p. 15.

6.7   Robert H. Fickies, Geological Survey, State of New York.

6.8   Robert H. Fickies, Geological Survey, State of New York.

6.9   Robert H. Fickies, Geological Survey, State of New York.

6.10   Robert H. Fickies, Geological Survey, State of New York.

6.11   Robert H. Fickies, Geological Survey, State of New York.

6.12   West Virginia Geological and Economic Survey.

6.13   West Virginia Geological and Economic Survey.

6.14   Dick Rader. Courtesy of New Jersey Geological Survey.

6.15   *Coasts in Crisis*, USGS Circular, 1990, p. 9.

6.16   *Geology of New York*, New York State Geological Survey, 1991, p. 234.

6.17   *Geology of New York*, New York State Geological Survey, 1991, p. 235.

6.18   Robert H. Fickies, Geological Survey, State of New York.

6.19   Geological Society of America, *Geology of North America*, Vol. A, GSA, p. 562.

6.20   Geological Society of America, *Geology of North America*, Vol. A, GSA, p. 562.

6.21   American Institute of Professional Geologists, Arvada, Colorado.

6.22   American Institute of Professional Geologists, Arvada, Colorado.

7.1   American Institute of Professional Geologists, Arvada, Colorado.

7.2   Donald F. Oltz, Geological Survey of Alabama.

7.3   American Institute of Professional Geologists, Arvada, Colorado.

7.4   Thomas Scott. Courtesy of Department of Environmental Protection, Florida.

7.5   Jonathan D. Arthur. Courtesy of Department of Environmental Protection, Florida.

7.6   *USGS Yearbook*, 1992.

7.7   American Institute of Professional Geologists, Arvada, Colorado.

7.8   *USGS Yearbook*, 1992.

7.9   American Institute of Professional Geologists, Arvada, Colorado.

7.10   *Georgia's Environment 99*, Environmental Protection Division, 11.

7.11   *Georgia's Environment 99*, Environmental Protection Division, 11.

7.12   American Institute of Professional Geologists, Arvada, Colorado.

7.13   Michael B. E. Bograd, Office of Geology, State of Mississippi.

7.14   *Coasts in Crisis*, USGS Circular, 1990, p. 8.

7.15   American Institute of Professional Geologists, Arvada, Colorado.

7.16   Michael B. E. Bograd, Office of Geology, State of Mississippi.

8.1   Indiana Geological Survey.

8.2   Indiana Geological Survey.

8.3   Robert H. Fickies, Geological Survey, State of New York.

8.4   Robert H. Fickies, Geological Survey, State of New York.

8.5   Robert H. Fickies, Geological Survey, State of New York.

8.6   *USGS Yearbook*, 1998.

8.7   American Institute of Professional Geologists, Arvada, Colorado.

8.8   Marvin B. Berwind, State of Tennessee.

8.9   Marvin B. Berwind, State of Tennessee.

8.10   Marvin B. Berwind, State of Tennessee.

9.1   Angela Braden, Arkansas Geological Commission.

9.2   Geological Survey of Canada.

9.3   Idaho Bureau of Disaster Services Archive.

9.4   Idaho Bureau of Disaster Services Archive.

9.5   Kenneth S. Johnson, Oklahoma Geological Survey.

9.6   *USGS Yearbook*, 1995.

9.7   Sigrid Cliff, Bureau of Economic Geology, University of Texas, Austin.

9.8   Sigrid Cliff, Bureau of Economic Geology, University of Texas, Austin.

9.9   Sigrid Cliff, Bureau of Economic Geology, University of Texas, Austin.

9.10   *Atlas of the United States*, Department of the Interior, 1970.

10.1   Denver, Colorado, Convention and Visitors Bureau.

10.2   Chicago, Illinois, Convention and Visitors Bureau.

10.3   Washington, D.C., Convention and Visitors Bureau.

INDEX

About the Author

ANGUS M. GUNN is Professor Emeritus, University of British Columbia. He is the author of *Patterns in World Geography*, *Habitat*, and *Heartland and Hinterland*.